OUR LIVING WORLD OF NATURE

The Life of the Pond

WILLIAM H. AMOS

Published in cooperation with
The World Book Encyclopedia

McGraw-Hill Book Company
NEW YORK   TORONTO   LONDON

WILLIAM H. AMOS *was raised in the Orient, where he first was introduced to aquatic and marine life in the Philippine Islands and in Japan. He has been associated with the New York Zoological Society, the Mt. Desert Biological Laboratory, and a number of marine laboratories in the United States and abroad, and he was a member of the Smithsonian-Bredin Expedition to the Lesser Antilles a few years ago. Mr. Amos is Chairman of the Science Department of St. Andrew's School in Middletown, Delaware, and is a Research Associate of the University of Delaware, Department of Biological Sciences. He has been a Senior Visiting Investigator at the Systematics-Ecology Program at the Marine Biological Laboratory in Woods Hole, Massachusetts, where he also serves as a consultant in biophotography. His work in fresh-water science, or limnology, commenced at Rutgers University and has continued during most of his professional life. Mr. Amos's present laboratory is on the shore of a highly productive pond, which he has studied for twenty years. His other major interests involve marine organisms and those of coastal estuaries. He is the author of many articles and books on marine and aquatic biology, most of which have been illustrated with his own biophotographs.*

Library of Congress Catalog Card Number: 67–16306

67890 NR 721

ISBN 07-001586-4

OUR LIVING WORLD OF NATURE

Science Editor

RICHARD B. FISCHER *Cornell University*

Board of Consultants

ROLAND CLEMENT *National Audubon Society*

C. GORDON FREDINE *National Park Service, The United States Department of the Interior*

WILLIAM H. NAULT *Field Enterprises Educational Corporation*

BENJAMIN NICHOLS *Cornell University*

EUGENE P. ODUM *University of Georgia*

HENRY J. OOSTING *Duke University*

OLIN SEWALL PETTINGILL, JR. *Cornell University*

DAVID PIMENTEL *Cornell University*

PAUL B. SEARS *Yale University*

ROBERT L. USINGER *University of California*

Readability Consultant

JOSEPHINE PIEKARZ IVES *New York University*

Special Consultants for The Life of the Pond

DALE F. BRAY *University of Delaware*

EDWIN T. MOUL *Rutgers University*

ROBERT W. PENNAK *University of Colorado*

Contents

THE POND WORLD   9

What is a pond? 10; What is the history of your pond? 14; A pond's many habitats 20; Plants are food factories 23; Consumers and decomposers 24; A multitude of plants 25; The transfer of energy 26; Pond succession 28; In the beginning 29; The young pond 33; A pond in its prime 34; The littoral zone 34; Away from the shore 41; Plants that float 44; Plants beneath the surface 46; The zonation of animals 50; The aging pond 51; The death of a pond 55

SEASONS AND DAYS   57

The coming of spring 58; The dragonflies 62; Creatures of the late spring 65; Early days of summer 66; Birds of the pond 72; The pond at dusk 76; The pond at night 77; A new day 79; The threat of drying 79; The waning of summer 80; A pond loses food 82; How food is restored 83; The chill of fall 88; To migrate or to sleep 89; The icy grip of winter 95

LIVING IN WATER   99

Plants that live in water 102; Plants and more plants 103; Flowers in the water 105; The simpler plants 106; The smallest of all 106; Animals that take to water 108; To swim like a fish 114; The diving beetle 116; Other insect swimmers 118; Jellyfish and crustaceans 120; Microscopic swimmers 122; Take a breath 123; How beetles breathe 124; Physical gills 124; Living gills 127; How crayfish breathe 128; Breathing without gills 129; A fish pump 131; Diving birds 132; A hungry multitude 133; The catchers and the eaten 138; The bullfrog 141; Underwater hunters 145; Insect predators 146; An eight-eyed wonder 149; To live with another 152; Parasites 153; A sanitary service 157

WORLDS WITHIN WORLDS 161

Life under the duckweed 162; A deathtrap for a home 164;
A floating home 166; Under a lily pad 167; Living on a
stem 168; The surface is a ceiling 170; The water strider 172;
Insect speedboats 174; Breaking through the barrier 176;
The depths of the pond 177; The burrowers of the bottom 178;
More bottom dwellers 179; Animal strainers 181; The crowded
world of the plankton 183; The tiny drifting plants 184;
Plant geometry 184; Plants that swim 185; A living globe 191;
The world of zooplankton 192; The smallest shrimps 193;
Other members of the plankton 195; The hidden life of the
soil 196; Animals among the sand grains 196; Exploring the
pond world 199

APPENDIX

Ponds in Our National Parks and Wildlife Refuges 202; What a
Fish Sees 206; A Guide to Some Common Pond Animals 208;
How to Learn More About a Pond 210; Homemade Ponds 214;
Keeping Pond Animals in the Home 216; Exploring the Microscopic
Pond World 219; An Invitation to Adventure 221

Glossary 222
Bibliography 227
Illustration Credits and Acknowledgments 228
Index 229

the different seasons. You should also take into account its geographical location, that is, its latitude and its altitude above sea level. All these factors affect a pond and its life.

Then you will want to find out the history of your pond. It is valuable to know something of the geology of the land around it, how the pond originated, and how long it has been in existence. There are several types of natural ponds, although in the United States natural ponds are no longer nearly so numerous as those built by man. In the mountains, landslides may fill stream beds with earth and rock, causing ponds to form. Erosion usually tends to destroy lakes and ponds, but sometimes it creates depressions that fill with water. In lowland river valleys, streams twist back upon themselves, forming *meanders*. If a meander loops too much, the river may eventually by-pass it. The isolated U-shaped body of water that results is known as an oxbow lake, but it is really a kind of pond, for it is usually shallow and quickly fills with aquatic vegetation.

Dune ponds, such as this one on Cape Cod, form where winds scoop out the sand of an ocean beach or lake shore and heap it in barrier dunes. Even though such a pond may be but a few hundred yards away from the ocean, its water may be perfectly fresh.

In the United States today, man-made ponds outnumber those created by nature. This farm pond, bulldozed out of a pasture, serves a variety of purposes: livestock drink from it; it acts as a reservoir that can be tapped for irrigation or fire fighting; it affords human recreation in the form of swimming, fishing, and skating.

Natural ponds often form near the deltas of rivers when silt deposits become high enough to dam off parts of the stream. They may also form behind barrier dunes on ocean beaches; the water contained here is as fresh as that farther inland and is not affected by the salt water just a few hundred yards away. Still another kind of natural pond may develop in cavities in limestone called *sinkholes* or *solution basins*. Some ponds are the last watery vestiges of what once were large lakes.

Glacial ponds, potholes, and kettles were once the most numerous, but now by far the greatest number of ponds are the result of man's activities, either planned or accidental. They may be ditches or canals, small reservoirs, millponds, quarry ponds, fishponds, or watering holes for cattle. Man-made ponds are used to supply water for irrigation, for fire protection, and for ice production. They may also be used to attract wild waterfowl, to control erosion, or to provide an area for recreation. Once established, each of these ponds will support a wide variety of life, both plant and animal.

Then, of course, there are also beaver ponds. Beavers have been constructing dams far longer than men have, and there is evidence that their ponds existed on this continent long before the dawn of recorded history.

NATURE'S MASTER POND ENGINEERS

Man's closest rival in his ability to alter his surroundings is the beaver. This big rodent requires a pond in which it can build its home and store its winter food—and where no pond exists, the beaver is well equipped to construct one by damming the waters of a stream.

The beaver's construction materials include timber, rocks, mud, and grasses; its most important tools are its chisellike front teeth, which can bring down a six-inch-thick tree in ten minutes. Guided by inherited behavior patterns that might easily be mistaken for a reasoning intelligence, a colony of beavers—usually no more than a dozen animals—can build and maintain a remarkably well engineered dam, often impounding dozens of acres of water. Beaver dams average about seventy-five feet in length, but many are much longer. (The record appears to be a prodigious New Hampshire dam three-quarters of a mile long, the joint effort of a number of beaver colonies.)

Beaver ponds such as this were once a common sight throughout nearly the whole of the North American continent, but trappers, in quest of the animal's highly prized pelt, nearly drove the beaver into extinction. Protective legislation was enacted in time to save the beaver, and the creature has made a strong comeback over much of its original range.

17

PONDWEED

HORNWORT

WATERWEED

overturn or submerge one of these leaves.

Mallards and pied-billed grebes swim and feed among the floating plants. Beneath them, pickerel lurk quietly; sunfish, perch, speckled bullheads, and chubsuckers swim about feeding, and perhaps being fed upon by larger consumers.

Wave action is reduced by the floating leaves and their slender stems, but not eliminated. When a wind blows, the leaves rise and fall gently with the small waves. Occasionally they are jerked suddenly as a turtle thrusts its way among the stems. Beyond the floating plants there is nothing but open water. Out there, in the limnetic zone, are the plankton feeders, such as shiners, and larger predacious fish. Mergansers swim low in the water, diving frequently beneath the surface to capture fish, frogs, and fresh-water mussels.

Plants beneath the surface

Out where the wind raises wavelets on a pond surface, there still is plant growth, but most of it is well below the surface. The plants that grow there seldom reach the air, except to produce flowers. Some submerged plants are seed producers and are quite complex; others are very primitive forms that reproduce by means other than seeds. Some of the pondweeds, particularly the crimped-leaf pondweed, grow in great beds and apparently choke out other bottom-dwelling plants. Where these weeds do not dominate, other submergents will appear: water celery, water milfoil, waterweed, fanwort, hornwort, and bladderwort, to mention only a few. How extensive the zone of submerged plants in a pond is depends almost entirely on how transparent the water is. If plankton is densely concentrated, or if the water is constantly opaque with suspended silt particles, then light cannot filter far down, and the zone where photosynthesis can take place is necessarily a shallow one. In some ponds, however, the water is clear, and green plants can be found growing fifteen or twenty feet below the surface.

Long strands of water milfoil reach upward from the pond's bottom toward the life-sustaining sunlight at the surface. Milfoil, whose name means "thousand leaves," is among the most common of *submergents*—plants that grow completely underwater, except for aerial flower stalks.

46

RICH WATERS AND POOR

For a variety of reasons, physical and chemical, ponds contain differing quantities of the materials necessary for life, such as oxygen, carbon dioxide, and various compounds of nitrogen, calcium, potassium, and phosphorus. Ponds that are rich in these materials are designated as *eutrophic* ("richly nourishing") by biologists, and those that are poor in them are designated as *oligotrophic* ("scantily nourishing"). The pond shown at the left is eutrophic: note the crowded zones of littoral vegetation and, in the underwater view, the relative murkiness of the water, indicating a teeming plankton community. The pond on the right, on the other hand, is oligotrophic: its shore is virtually free of emergent vegetation, and its water is so devoid of plankton that sunlight penetrates all the way to the bottom, where only a few submergent plants struggle to live in the wet wasteland. In general, warm, shallow, southern ponds tend to be eutrophic; cold, deep, northern ponds are likely to be oligotrophic.

cabbages unfold their flowers and then their leaves from tight colorful buds that pushed up from the damp ground weeks earlier. Buds on many emergent plants open to reveal small green shoots that lengthen with every passing day. In quiet coves, masses of green algae rise to the water's surface, buoyed by oxygen bubbles given off as by-products of photosynthesis. A few duckweed and watermeal plants that survived the winter dot the surface.

During the first bright days of spring the winter plankton still remains, but within a week or so there are signs of the great plankton bloom that is to come: phytoplankton increases—diatoms, flagellates, and desmids. In shallow water along the shores, the eggs of rotifers and water fleas hatch, adding to the zooplankton. Water fleas and copepod crustaceans reproduce in great numbers.

Many of the insect larvae that passed the winter in a nearly dormant state now become hungry and active. Dragonfly and damselfly nymphs stalk across the bottom on the dead remains of last year's vegetation, seeking prey which they capture with hinged, expansible jaws that shoot out rapidly before the victim can escape. Amphipods and isopods that swam slowly about beneath the ice during the winter now go about their activities in a more frenzied manner. On a sunny day, among the dead and dried stalks of old emergent plants, you can find water striders skimming the surface. Water beetles make their appearance, but not yet in large numbers.

Western skunk cabbage differs from its eastern counterpart in possessing bright yellow flower hoods. The two plants are not close relatives, but they are similar in habit and habitat. The odor they emit, so offensive to the human nose, is nevertheless highly attractive to certain species of flies that move from plant to plant and carry out the vital task of pollination.

59

Hooded mergansers frequent secluded woodland ponds, where they nest in hollow trees, in stumps, or underneath overhanging banks. The birds are easily identified by the male's fan-shaped black and white crest, which can be raised or lowered and which plays an important role in his colorful courtship ritual. Less vegetarian than many ducks, mergansers include quantities of fish, frogs, and other animal foods in their diet. Because of the resultant "gaminess" of their flesh, the birds are molested by human hunters to a lesser degree than some of their more herbivorous, and therefore more tasty, relatives.

In its threat posture (*top*), a dragonfly flicks its wings open and elevates its abdomen about forty-five degrees from the surface on which it is resting. On warm days, a resting dragonfly may use its wings to shade its abdomen (*center*). On very hot days, the insect shields its thorax with its wings and points its abdomen upward (*bottom*) so that it receives as little direct sunlight as possible.

Soon the first waterfowl of the season arrive from the South. You can see common mergansers, ring-necked ducks, and pied-billed grebes diving deep into the pond's cold waters. And on a bright day you may also glimpse a muskrat swimming through the dry reeds that rattle in the chill wind.

Under the warming sun certain tiny flies known as midges emerge from their pupal cases at the surface, take wing, and mate hovering over the pond. These are the first of the pond's insects to fly in the clear spring air, but their adult lives are short and are often over before the cold night comes.

The bass and sunfish that spent the winter in deeper water now come close to shore to feed on the tiny organisms that appear with every new day. On warm days spring peepers call from marshy spots deep in a wooded cove. A few frogs appear in shallow water, but they are quiet.

As May approaches, the pond bursts with life. The warming sun has raised the temperature of the surface water several degrees. Two or three species of pond snails join the amphipods and isopods. Planarian flatworms begin depositing their stalked cocoons on pebbles and submerged sticks; after only a week or two, several tiny, almost colorless planarians crawl out of the ruptured cocoons and begin their life on the bottom of the pond.

The dragonflies

Dragonflies now make their appearance, having crawled as nymphs onto plant stems above the surface. By mid-May there are a great many of them darting above the pond. They capture emerging and flying insects by grasping them with their legs, which form a kind of basket.

You can learn a great deal about dragonflies simply by sitting quietly by a pond. Perhaps a dragonfly will rest for a while on a bare twig near you and watch for passing insects. You will then be able to see its extremely large eyes and mobile neck. Every so often, its head will twist about quickly as it sights some passing prey. Most of the dragon-

This dragonfly spent the first year of its life as a nymph, an active predator of the pond bottom. Now it has crawled up a plant stem, molted its final nymphal skin, and emerged as a winged adult, one of the most superbly skillful fliers of the insect world.

62

HONKERS OF THE POND

Of the many water birds that use our northern ponds as summer nurseries, none is more widely known than the big, stately Canada goose. Ready to breed at the age of two or three years, goose and gander mate for life. A nest of sticks and grass is built—frequently atop a beaver lodge—and the female lays from five to nine buff-colored eggs, which she incubates for about a month. During this time, the gander vigilantly guards his nesting mate; a human venturing too near the nest site is almost certain to be driven away by a furious attack from the big male bird. Two days after hatching, the downy yellow goslings are led down to the water by both parents, who continue to care for their offspring throughout the summer. When the days shorten, the Canada geese leave their breeding ponds and head south again.

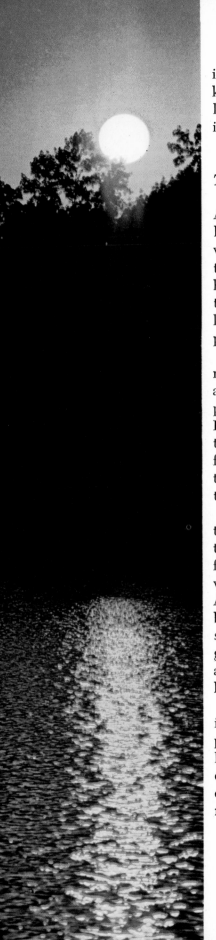

insects, frogs, mice, crayfish, and even berries. Certainly kingfishers do more good than harm. A decline in the population of pond fish can nearly always be traced to a decreasing food supply, and very seldom to large predators.

The pond at dusk

A pond in summer is a fascinating place. You can learn a lot by sitting down next to the water's edge and watching what goes on around you. Probably you will not have the time, or the patience, to spend twenty-four consecutive hours there, but it would be well worth your while to visit the pond at intervals during the day and night. The pace of life in and around the pond constantly changes as the hours pass.

At dusk, water lilies have already closed their petals, and many animals are quiet. Some creatures, however, are most active during the hours just preceding darkness. Yellow perch feed busily now, and midges dance over the water. Dozens of swallows skim the quiet water, catching some of the midges and other flying insects that have just transformed from aquatic nymphs and pupae. Every now and then a swallow will furrow the surface as it scoops up water to drink.

It is during daylight hours that most midges and mosquitoes emerge from their aquatic pupae to become adults, though a few seem to prefer early evening. The whole transformation takes only a matter of seconds. First you see a wriggling pupa rise to the surface from the muddy bottom. As soon as the pupa hits the surface film, it splits down the back. The adult midge pops out, stands on the empty pupa skin for a moment, and then takes wing. During the emergence, the pupal case often darts across the water. This action may be caused by a secreted substance and may help the adult to emerge and dry off its damp folded wings.

As darkness falls, silent winged creatures follow strange irregular courses over the pond; they are little brown bats pursuing moths, midges, and other flying insects, which they locate by means of their remarkable sonar system. A few of the moths are able to "tune in" on the bats' ultrasonic cries, and their evasive actions make the aerial chases even more erratic.

76

The pond at night

With darkness, the pace of life in the pond slackens. Phytoplankton and shoreline plants no longer manufacture food and release oxygen; instead they consume oxygen and use it to extract energy from stored food. A barred owl hoots along the shore, but most birds are sleeping. If you move slowly and quietly to the water's edge and shine a flashlight into shallow water, you may see fish resting almost motionless on the bottom. Only their gently waving fins betray their presence. Since fish have no eyelids to close, they may seem to be awake, but they are not.

Frogs are active in the early evening; then, as night wears on, they call less and finally fall silent.

Most aquatic insects and crustaceans seem to be as active at night as they are in the daytime. Indeed, because you can see them easily with a flashlight at night, they appear to be even more abundant and more active. Amphipods dash from under leaves and fallen limbs, swimming rapidly on their sides. Below them, the slower isopods crawl over decayed vegetation, scavenging bits of organic matter.

Try hanging over the edge of a dock or over the side of a boat at night, with your face close to the water surface. Shine your flashlight straight down and look for the individual members of the zooplankton twisting and turning through the water. If your eyes are good and you look closely, you will be able to distinguish various animals: copepods swimming with skips and stops, water fleas swimming jerkily, ostracods swimming rapidly close to the bottom, water mites swimming smoothly over the bottom, large rotifers swimming smoothly near the surface. If you are very attentive, you may even see one of the larger protozoans, *Paramecium*, swimming evenly in a spiral path.

Once you get used to looking at organisms of this size, you are startled when a relatively huge water beetle comes dashing along. Then you realize that from the point of view of a nearly microscopic planktonic creature, this insignificant pond is a vast world indeed. Being so large ourselves, we tend to be aware of only the largest creatures. If the sizes of typical adults of every one of the world's animal species were tabulated, the average probably would be about that of a house fly. And, of course, the average size of all pond-dwelling species would be very much smaller.

77

A new day

As dawn approaches, yellow perch start feeding actively again. Water boatmen, aquatic beetles, leeches, and planarian flatworms move about more. Later, when the sun shines directly on the pond, many of these creatures tend to congregate under submerged logs and stones, and in the shade of vegetation. (Because they move slowly in the dark and rapidly in the light, they inevitably collect in dark, protected places.) Frogs and turtles become more active now that they can use their eyes for hunting.

Aquatic animals react to the sun's warmth in various ways. Many of the larger ones seek the shade of floating leaves or submerged objects. Turtles and snakes emerge from the water and bask in the sun.

As the water becomes warmer, it holds less dissolved oxygen, and some animals begin to have difficulty breathing. Tube-dwelling midge larvae and burrowing *Tubifex* worms increase their waving and thrashing movements, thus bringing more water their way. The lower the oxygen content of the water becomes, the farther they stick out of their tubes and the faster they wave. Fish rise to the surface to breathe the oxygen diffusing into the water from the air. Some insects must make frequent trips to the surface to replenish bubbles of air that they carry underwater. Damselfly nymphs climb up the stems of emergent plants and place their leaflike abdominal gills against the surface film.

By late afternoon the water temperature begins to fall, breathing becomes easier, and the pace of life in the pond becomes less frantic. As dusk approaches, many animals begin to feed more actively.

The threat of drying

Summer is not always a time of plenty for pond animals. It can be a time of death or dormancy, when all the water in a shallow pond evaporates to reveal a sun-baked brick-

The early morning mist drifting across the water marks the start of a new day in the pond. Some creatures are now stirring into activity, and others are retiring to hideouts to await the return of darkness.

79

These yellowing water lily leaves embedded in cracked mud once floated on the surface of a shallow pond. The hot summer sun can reduce a flourishing pond to a bone-dry wasteland. In the face of such a catastrophe, some of the pond's residents migrate to other waters. Others, less mobile, either perish or go into a state of suspended animation to await the replenishing rains of autumn.

hard bottom. It seems improbable, but many of the pond's organisms survive drying and reappear after the autumn rains and winter snows. Bacteria and protozoans survive long droughts in microscopic spores and cysts. Snails, pill clams, water fleas, copepods, ostracods, leeches, and a variety of other purely aquatic animals enter a state of suspended animation, or *estivation*, as the mud hardens around them. Certain ostracods and water fleas have been known to estivate for twenty years, and fairy shrimp may estivate for even longer periods.

Most insects are not bothered particularly by a drying pond, although those that are entirely aquatic in their habits during their nymphal and larval stages may die if they cannot complete their development before all the water evaporates. A few kinds of mosquitoes actually have to complete their life cycles in temporary ponds. They usually deposit their eggs in dry depressions; after rain fills these puddles or little ponds, larvae emerge and begin their active lives.

Larger adult aquatic insects easily survive droughts, for most of them are capable of migrating elsewhere. Frogs and turtles, too, can seek more favorable habitats. Among the larger animals, only fish are unable to escape a drying pond.

The waning of summer

By July the larger organisms of the pond are fully developed and in their prime. Phytoplankton, however, is no longer multiplying as fast as it did in spring, and so there are fewer of the small creatures that feed on it. The animal life that is so evident in the pond now is composed largely of filter feeders, scavengers, and several levels of predatory consumers feeding upon an enormous amount of animal substance that got its start with the primary production by plants earlier in the year.

Shoreline plants continue to grow and to flourish. Almost all summer long, flowers along the shore lend their color and fragrance to the pond. They attract pollinating insects, some of which are caught by frogs, dragonflies, and other predators that lurk in the dense masses of emergent plants.

As summer passes, the water becomes dark with increased quantities of zooplankton and organic particles, and perhaps with soil particles washed from land during summer

80

rains. You seldom see the larger animals that were so evident in early summer; they have laid their eggs and raised their broods. Frogs no longer call to one another. The markings of sunfish, shiners, and other fishes are not so vivid, and the colors of the smaller birds, such as redwings, do not seem so brilliant as they did in May and June. The leaves of spatterdock take on a battered, dry appearance, and a few of the emergent plants begin to wither in the August heat.

As September approaches, dodder flowers along the shore, conspicuous wherever it grows. It is one of the few parasitic seed plants, and it often attacks water willow, a common emergent. Vinelike, it winds its slender yellow stems around the stems of the water willow.

Dodder has no leaves to speak of, and it does not manufacture food by photosynthesis; instead it obtains nourishment from a host plant. Budding from the dodder stems are tiny extensions that penetrate the tough tissues of the water willow to the depth of the conducting vessels. With these stem extensions dodder removes food manufactured in the leaves of the host. Apparently it does the water willow little harm. The whole shoreline may appear to be covered by heavy yellow cobwebs in late summer, but there is always just as much water willow the next year.

The aquatic insect population is at its peak in late summer and early fall. Many of the insects that develop in water and leave as flying adults emerge at this time of year, occasionally filling the air in swarms. The emergence of mayflies earlier in the season is a remarkable phenomenon in some areas of the country, but it is of short duration. More flying insects leave the pond in the late summer than at any other time of year.

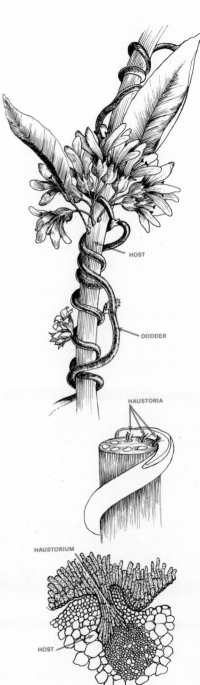

One of the strangest of the pond's shoreline plants is dodder, a leafless parasitic plant that cannot manufacture food of its own. Instead, dodder winds its way along the stems of such plants as water willow and sends out specialized structures called *haustoria*, which penetrate into the host plant and drain off a portion of its food substances. In July and August, dodder produces dense clusters of tiny whitish flowers. Dodder is an annual and dies in the fall; its seeds fall to the ground and pass the winter at the base of the host plant, where they germinate with the return of warm weather. Once the upward-groping seedlings have gained a grip on a host plant, they entirely break off contact with the soil.

81

A pond can be thought of as a sort of chemical factory receiving energy and raw materials from the outside world and, through a complex set of internal operations, manufacturing a variety of products, some of which return to the outside world. The principal source of energy is the sun, without whose light and heat the pond's "machinery" could not function. Raw materials come from a variety of sources: chemicals leach into the water from the soil (although the reverse may also occur); rain also washes in sediment and other materials from the land; the air furnishes oxygen and carbon dioxide; birds, mammals, and reptiles contribute their waste products and, sometimes, their dead bodies; leaves, pollen, and insects fall onto the surface and eventually are consumed; also a . . .

A pond loses food

The organisms that depart from a pond take away reserves of energy and nutrients contained in the organic matter in their bodies. Midges, for example, may be very small insects, but the millions that leave even a small pond during the course of a summer account for a considerable loss. And most departing insects are much larger than midges.

Larger animals, too, take organic material away from the pond. Frogs spend their tadpole days as primary consumers, browsing on algal coatings on plant stems and rock surfaces; but they grow quickly, leave the pond, and spend much of their time on land nearby. As long as waterfowl are present, they continually take organic matter from the pond; so do herons, ospreys, and kingfishers. Otters, muskrats, and raccoons regularly remove plant and animal food from the water and the shores.

A pond is a kind of sieve through which water passes

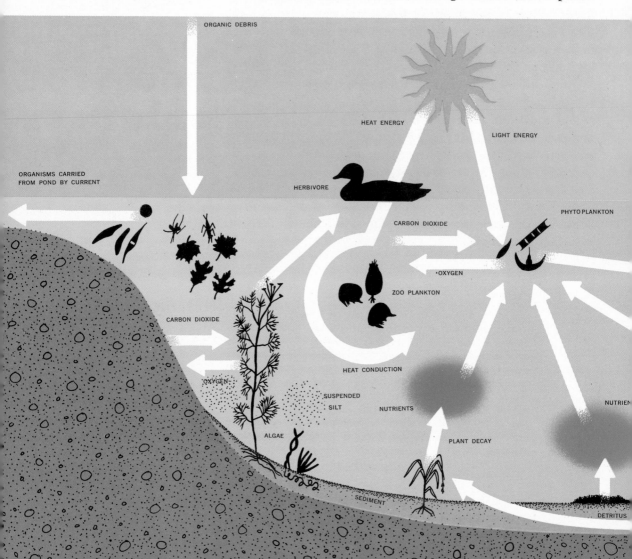

frogs contribute a little. Creatures that lay their eggs in the pond add to its organic stores, but only to a slight degree, for the eggs are minute. The most important source of food for the pond community is the primary production of its green plants.

The chill of fall

By early fall the countryside around the pond is a riot of color. People take walks to admire the fall foliage, but how many notice the colors of the smaller plants of a pond? The delicate pink clusters of water smartweed flowers cover the advancing front of emergent vegetation; clusters of water willow flowers form islands of purple; the deep red flowers of spiked loosestrife rise above the low-lying emergents.

Beneath the pond's surface, the fresh-water sponges look sickly. During the winter they will disintegrate, but their cells already have produced thousands of tiny *gemmules*, protected sporelike bodies that can withstand the rigors of winter and start new colonies in the spring. *Pectinatella*, a bryozoan, or "moss animal," that lives in large gelatinous colonies, solves the problem of surviving in winter in a similar way, by producing tiny disk-shaped objects known as *statoblasts*. They become entangled in vegetation close to the surface, favorable places for next year's colonies to begin growing.

During the summer, *Hydra*, a little tentacled creature that feeds on zooplankton, reproduced mostly by budding new individuals from its trunk; but now it reproduces sexually, producing a few thick-walled eggs. Female rotifers are beginning to lay special winter eggs, after having produced great numbers of thin-walled summer eggs that invariably turned into more females, exact copies of their mothers. Many of the rotifers die, and certain species survive only in the safe winter eggs.

By now, spatterdock and other pond lilies are falling to pieces. Their fragments gradually filter down to the bottom and add to the organic blanket coating the mud. Leaves from the emergents blow off, and the stems dry out and wither, leaving only the rootstocks on low, protected stems to survive the winter. With every gust of wind, multitudes of

GEMMULE
(*Spongilla lacustris*)

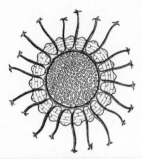

STATOBLAST
(*Cristatella mucedo*)

THECA
(*Hydra littoralis*)

88

leaves from the trees fringing the shoreline drift down into the water and are caught there. Some blow like miniature boats across the surface, but eventually they sink and become part of the rich sediment on the pond bottom.

The sun sinks a little lower in the sky each day, and the water cools, but not enough yet to hinder the growth of plankton. With so many nutrients entering the pond there may be another bloom of phytoplankton before the year is over, with a corresponding increase in zooplankton. But this late bloom will not equal that of the spring.

To migrate or to sleep

Suddenly you are aware that there are no more frogs about, and very few turtles. Only on sunny fall days do you find occasional painted turtles basking on shoreline logs, and with the first chill days they, too, are gone.

Turtles, snakes, frogs, and salamanders are not able to regulate their body temperatures like birds and mammals. They cannot cope with freezing temperatures and must hibernate. Water snakes generally find protected spots in old logs and under stones along the shore, and so do many of the salamanders. Their bodily processes slow down, and they become totally inactive. Frogs and turtles spend the winter under the mud at the bottom of a pond. Frogs have many blood vessels under their loose, moist skin, and while they hibernate they are able to absorb all the oxygen they need directly from the water. Turtles, with their watertight skin, have another means of obtaining oxygen. They inhale and exhale water through a large posterior opening, which serves also as an outlet for their excretory and reproductive systems. This cavity is extensively supplied with blood vessels and acts as a kind of gill.

The various ways in which plants survive the winter are less complex than those of the animals, but just as effective. Though the large leaves of water lilies die, the sizable stems and root systems live on in the bottom. This is the case with most emergents. Submerged plants such as waterweed and bladderwort produce either seeds or protected sprouts, which live through the winter after the rest of the plant has died. Algae remain somewhat active all year, even under the ice, or spend the winter as spores.

Perhaps more than most other aquatic habitats, the pond is responsive to the shifting seasons. The pond world is a world of extremes, and the plants and animals that populate it must endure these extremes or perish. In the spring, when conditions favor growth and proliferation, these activities are carried on with furious intensity. Summer may be a period of relatively easy living—or it may bring the threat of death through drought. Fall allows a brief renewed frenzy of growth in preparation for winter; winter brings the harshest trials of all. Yet despite the perils, the pond's community of living things thrives in wondrous variety and abundance.

The next four pages show a New Jersey pond during the four seasons: spring, summer, fall, and winter.

89

lives in wet soil along the shores of ponds and marshes. Its burrows may extend out under a pond and actually lead into it. A star-nosed mole is a fine swimmer and preys upon small fish and also crayfish, aquatic insects, and a variety of other pond invertebrates. It is one of the few mammals that is active in the water all year, and it even swims under the ice in search of food.

Although the pied-billed grebe inhabits ponds throughout North America, you are unlikely to see one at close range. It is a timid bird, and dives or ducks under the water instantly, long before an unsuspecting human visitor has a chance to see what it is. The grebe spends much of its time in the heavy vegetation at the edge of the pond, where it feeds on small fish and other aquatic animals. It swims effortlessly underwater, pursuing its prey, for long periods of time. On land it is helpless and cannot take to the air, as it does from water, because its legs and webbed feet are placed far back on its body.

Nearly all the turtles that live in a pond have very large webbed hind feet, which are efficient paddles for swim-

Sleek, playful, and incomparably agile in the water, the otters are the most thoroughly aquatic of all the fresh-water mammals. They are confirmed meat eaters: muskrats, young beavers, fish, shellfish, frogs, turtles, snakes, earthworms, crayfish, and water birds are all part of their diet. Overhunting by fur trappers has greatly reduced our otter populations, and though otters are still widespread, they are now seldom seen.

ming. The front feet, with their long claws, are not heavily webbed and are used mainly for tearing apart food and crawling along the bottom.

These aquatic reptiles are more skillful in the water than their cumbersome shapes would suggest, for they are streamlined and muscular and are able to stay underwater for long periods of time. They are very timid and seldom allow you to approach them; a large red-bellied terrapin basking on a log will slip into the water when you are drifting silently in a canoe a hundred yards away. Some turtles, such as the musk turtle and the snapping turtle, seldom leave the water even to sun themselves, but females of all species climb the banks of the pond in the spring to lay their eggs in the soil.

There are only a few thoroughly aquatic amphibians in our ponds. Bullfrogs and green frogs are in the water most of the time, but the slender spotted newts, with their large finned tails, are even more committed to an aquatic life. Spotted newts go through a rather complex life cycle. Their eggs are fastened singly to stems and leaves of aquatic plants, and the larvae that emerge have external gills and develop as truly aquatic forms. Later, they lose their gills and crawl out on land, becoming for a while land forms known as red efts. When they mature, after two or three years, they return to the water, develop finned tails again, and live in the water from then on—but with lungs, not gills.

RED EFTS

LARVAE

SPOTTED NEWT

THE POND'S LIVING FOSSILS

Turtles are an exceedingly ancient group of animals, having evolved from a primitive land-dwelling reptile some 200 million years ago—long before the first dinosaurs roamed the earth. That the turtles have so successfully stood the test of time speaks well for their unique boxlike body armor, into which they retire when danger threatens. A variety of turtles make their homes in ponds. For the most part shy and mild-mannered, these placid creatures reach maturity in three to five years and may live for fifty years or even longer. Given intelligent care, many species make interesting and hardy pets.

Like many turtles, the red-bellied terrapin spends much of its time basking in the sun—often on a log or rock, but never far from deep water, into which it scrambles at the slightest disturbance. Sunbathing increases the rate of the turtles' metabolism, discourages the growth of algae on their shells, and helps control external parasites.

dreds of microscopic stinging capsules arranged in clusters. As the animal sinks, it paralyzes and collects planktonic animals that brush against the tentacles. When the jellyfish gets close to the bottom, it turns right side up and swims to the surface again.

Some animals migrate vertically through the water without swimming at all. Pond snails and *Hydra* secrete gas bubbles that carry them to the surface after they release their hold on the bottom. Once at the surface, these creatures crawl on the surface film, which to them is a firm ceiling.

Planktonic crustaceans, whether they are copepods or water fleas (cladocerans), generally swim by beating their legs or waving their antennae. They are so small that they displace very little water and produce very little friction as they push their way forward. As a result, it takes relatively little effort for them to move rapidly and for great distances through the water.

Rhythmic contractions of its thumbnail-sized body send this fresh-water jellyfish upward through the water by gentle jet propulsion (1) Reaching the surface (2), it turns over and sinks to the bottom again (3), with tentacles outstretched. Then it rights itself (4) and begins a new trip to the surface. It eats tiny creatures captured and killed by its stinging tentacles. Although jellyfish are abundant in the sea, the species shown here, *Craspedacusta*, is the sole member of the group to be found in the fresh waters of North America.

3

4

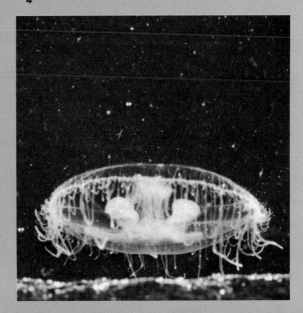

Microscopic swimmers

Among the thousands of species of single-celled protozoans, only three fundamentally different forms of locomotion have evolved. The ciliates, represented here by *Spirostomum* (*left panel*), propel themselves through the water by means of numerous hairlike organs called *cilia*. Flagellates, such as *Euglena* (*center panel*), have evolved a whiplike organ of locomotion called a *flagellum*. The flagellum functions as a sort of one-bladed propeller, with its looping undulations pulling the animal through the water in a characteristic spiral path. The third type of protozoan locomotion is exemplified by *Amoeba* (*right panel*), which moves by means of pseudopodia—literally "false feet." A pseudopodium is an extension of the animal's body, made possible because the cytoplasm, or cell contents, can exist in either of two states, one fluid and one jellylike.

The world of protozoans, or one-celled animals, is one that you cannot explore without a microscope. But if you are fortunate enough to have one, it is a world that will hold your attention for hours and perhaps for the rest of your life.

Flagellates use their long, whiplike flagella in several ways. Some species have flagella that extend forward and create vortexes which draw them forward. Other species have flagella that trail behind and whip back and forth, acting like sculling oars. An accessory flagellum may extend in a groove around the animal, causing it to spin on its axis.

Ciliates have many rows of small coordinated hairlike structures, or *cilia*, which beat together rhythmically. Within their microscopic bodies are complex patterns of coordinating fibers that control the beating. The ciliate is unable to swim properly if these fibers are cut with microsurgical tools. They do not constitute a nervous system, for the animal is little more than a single cell, but serve the same general function. In some ciliates, tufts of cilia are fused together and function as little tapering legs. Protozoans of

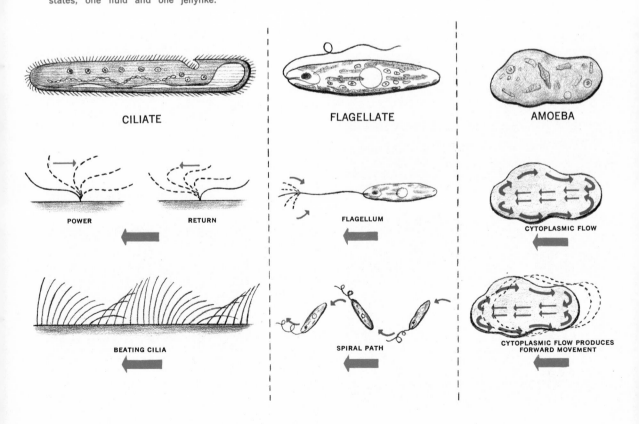

CILIATE   FLAGELLATE   AMOEBA

POWER   RETURN   FLAGELLUM   CYTOPLASMIC FLOW

BEATING CILIA   SPIRAL PATH   CYTOPLASMIC FLOW PRODUCES FORWARD MOVEMENT this type resemble scurrying mice as they creep through detritus in the field of a microscope.

Although the familiar *Amoeba* does not swim, some kinds of *Amoeba* can remain suspended in the water by forming long ray-shaped extensions of cytoplasm. A few related species that are frequently picked up in plankton tows, the "sun animals," or heliozoans, have raylike extensions of exceptional beauty.

Take a breath

Nearly all living cells require oxygen in order to release the energy locked in food. The complicated process by which cells methodically break down and release energy from food is known as *cellular respiration.* It is not the same process as breathing, which we call "respiration." Breathing is simply the act of acquiring oxygen from the air by means of a functional device such as a lung. As we have seen, pond plants may store some of the oxygen of photosynthesis in the airspaces of their stems during the day and then draw upon these reserves at night. Pond animals, on the other hand, do not produce any oxygen of their own, but they have developed some unusual ways of obtaining it.

Many creatures—turtles, frogs, immature and adult insects, and some snails—live more or less permanently in a pond but breathe atmospheric oxygen. Frogs have slightly elevated nostrils, and some turtles have elongated noses that poke out through the surface without any other part of the head showing. Snorkels of this sort allow an animal to remain submerged while it replenishes its air supply.

More extreme examples are found among insects. The water scorpion has a pair of posterior filaments which, when pressed together, form a breathing tube. Rat-tailed maggots, larval insects found in low-oxygen waters along pond margins, have telescopic breathing tubes of remarkable proportions; when fully extended, they can be four times the length of the insect's body. Mosquito larvae wriggle upward through the water to the surface, where they poke a small snorkel through the elastic film. The tip of this tube is equipped with tiny hairs that repel water and thus keep the tube open to the air for as long as the larva hangs there. Mosquito pupae have a pair of similar tubes that open from the enlarged midregion of the body, rather than from the tip of the abdomen as in the larvae.

How beetles breathe

When you sit by a pond and watch the activity of large aquatic insects, you will be impressed by the frequency with which many of these active creatures come to the surface. They dart up, hang head downward from the film for a few seconds, and then swim down and out of sight. Like all adult insects, each of these creatures has throughout its body a network of very fine tubules, the tracheal system. Air passes in and out of this system through a series of pores, or *spiracles*. The actual exchange of air is brought about by the expansion and contraction of the abdomen, rather like the action of bellows.

The spiracles of aquatic insects open either to a reserve of trapped air carried outside the insect's body or to the rear of the abdomen. *Dytiscus* is a good example of an insect that carries an air supply down with it. Its large, heavy wing covers shield air bubbles trapped underneath them. The insect draws upon this air as needed, but periodically it has to return to the surface to replenish the supply.

If you look at other adult insects underwater, or in an aquarium, you are sure to notice that they, too, carry air in some fashion—under wing covers, on the underside of the body, or perhaps encasing the entire body. While some may use the air directly, as large diving beetles do, others make use of it in a different way.

Physical gills

Land animals that have returned to an aquatic environment have not changed their lungs back to gills. Whales, sea lions, otters, turtles, and other aquatic mammals and reptiles all have to surface to breathe. It might be to their advantage if they could extract oxygen from water, but none of them can. One animal group now living in water but originally from the land—aquatic insects—has partially solved this problem.

Backswimmers, certain other diving bugs, and small diving beetles carry air bubbles underwater. These bubbles act as "physical gills" rather than as biological, or living, gills.

A backswimmer uses up the original oxygen supply, contained in the bubble, in a few minutes, yet it remains submerged and active for much longer periods of time. How?

The half-inch whirligig beetle is primarily an inhabitant of the surface film, but it occasionally dives, with a bubble of air trapped beneath its wing covers and partially protruding near the tip of its abdomen. At the surface the beetle obtains oxygen directly from the atmosphere, but when it is submerged the air bubble acts as a physical gill. As a larva the insect possesses true fleshy gills which extract oxygen from the water.

Oxygen from the surrounding water continually diffuses through the surface film of the bubble and maintains its contents at the normal atmospheric ratio of twenty percent oxygen and eighty percent nitrogen. A diving insect gets rid of carbon dioxide in just the reverse fashion: the insect simply exhales it into the bubble, and it readily diffuses into the water.

You might think that their physical gills would allow these small diving insects to remain permanently submerged, but such is not the case. Their bubbles gradually decrease in size as nitrogen escapes slowly into the water. So these insects do have to surface to replenish their bubbles, though much less frequently than the large diving beetles. Some adult insects, however, can stay underwater almost indefinitely. The water boatman, for example, uses the air between its wings and body so efficiently that it seldom comes to the surface. And many of the extremely small aquatic adult insects never come to the surface at all. Their bodies are covered with densely packed water-repellent hairs, two million to a square millimeter. Beneath the

When submerged, the backswimmer breathes by means of a physical gill, a bubble of air through which oxygen is absorbed from the surrounding water and waste carbon dioxide is discharged. Although the backswimmer must eventually return to the surface to renew the bubble, this unusual adaptation allows the insect to remain underwater for as long as six hours.

125

Deer, such as this white-tailed doe, are frequent visitors to ponds, not only to drink and to browse on spatterdock and other vegetation, but apparently also to enjoy a dip in the water. The best time to see these shy animals is early in the morning or just after sunset.

The catchers and the eaten

As you drift quietly in a boat across a pond, you may hear a rapid series of high, cheeping whistles far overhead. Looking up, you see a great brown and white hawk circling. It hovers, folds its wings, and dives straight down into a patch of yellow water lilies. It hits with a loud splash, throwing water into the air, and disappears for a moment. Then, shedding drops in a glistening spray and rising into the air with powerful beats of its great wings, the hawk—an osprey —reappears, clutching a bass in its talons. It flies directly to a dead tree overhanging the edge of the pond, shifts the fish about with its claws, and pecks at it. Then the bird takes off and flies out of sight, still holding the fish tightly. To watch the hunting of an osprey is a thrilling experience, and one you will not forget, for it is likely to be the most dramatic example of predation you will see in a pond.

Other hunters are not so easy to see. From a distance you may glimpse a rippling wake in the water and then a sudden churning turmoil. A few moments later a sleek otter emerges on the shore nearby and bounds away with a large glistening fish in its mouth. Later, as you sit rocking gently in your boat, you are alarmed by a thunderous snap and splash in the nearby lily pads. The water swirls, then quiets down, and all is as before. What was it? Probably a large snapping turtle crept along the bottom up to a basking frog or fish and then, with lightning speed, thrust out its long neck and clamped down its sharp jaws on the victim.

There is something rather impressive about a turtle weighing thirty, forty, or even as much as sixty or seventy pounds, but a pond usually cannot support more than a very few such giants. Snapping turtles occupy the peak in a pyramid of numbers and have a fairly limited food supply. Although they may occasionally drag ducks beneath the surface, waterfowl certainly do not commonly figure in their diets. They eat more plant material than anything else, followed by fish, dead organisms, and invertebrates.

The pond offers no more thrilling sight than that of an osprey hovering fifty feet above the surface, then plummeting talons-first into the water and coming up with a fat wriggling fish. Occasionally an osprey, strictly a fish eater, will drown through locking its talons into prey too large to lift from the water.

138

Because the bones of their jaws and skulls are joined by stretchable ligaments, snakes are able to swallow seemingly impossibly large prey, such as this bullfrog being consumed by a water snake. The snake's backward-pointing teeth assist in this process but are not used for chewing; the prey goes down whole and is dissolved by the snake's powerful digestive fluids.

The largest of American frogs, a full-grown bullfrog may have an eight-inch body and ten-inch hind legs. Anything that moves and is not too large to be stuffed into the bullfrog's ample mouth is fair game: insects like the dragonfly being eaten here, smaller frogs, whole crayfish, and even small birds and mammals. Bullfrogs have been known to live for fifteen years.

up its windpipe. Once you hear the scream, you will never forget it, especially if you hear it out by a pond on a dark night.

Bullfrogs are solitary animals and are more aquatic than most frogs, spending nearly all their time floating in the water and diving to the bottom. They do not inhabit every portion of a pond's shore, but seek areas overgrown with willows and other shrubs, where roots and stems form tangled mats.

Bullfrogs are not discriminating eaters. They reach with their muscular, elongated tongues for a variety of creatures —insects, small frogs, young snakes, small fish, indeed almost anything that moves and will fit into their mouths.

144

Underwater hunters

When you enter a bed of yellow water lilies or other plants with flexible emergent stems, sit quietly and wait for action. It may not be long in coming. Not far away the stems will begin to move suddenly as a fleeing fish brushes against them, taking the most direct route of escape. The passage of the fish is marked clearly by the moving, bending plants. Close behind comes a mighty rush, and the plants dance and sway as a larger and far more powerful creature surges through the water. You can only guess, but judging from the speed it displays, you decide that it must be an attacking pickerel. Although not related to the barracuda, the pickerel is similar in many ways to this great marine predator. Both hang motionless for long periods of time; both patrol definite regions; both are elongated and arrow-straight; both attack by sight with blinding speed; and both have sharply pointed teeth. A favorite pickerel locale is among emergent vegetation, especially in the deep shadows under lily pads.

Pickerels do not tolerate competition, and so where one is found, others will not be near. The larger pickerels have territories which they police, and in which they do most of their feeding. Anything that comes their way is fair prey; they will capture and devour ducklings, frogs, fish, small snakes, newly hatched turtles, insects, and crayfish.

Chain pickerel, grass (mud) pickerel, and redfin pickerel are the species common in ponds. Although far smaller than their relatives, the lake-dwelling northern pike or the great muskellunge, they are no less ferocious.

A streamlined pickerel watching for prey often hovers motionless just below the pond's surface, its dappled markings blending with the shadows of floating vegetation. The victim, seized in a lightning-fast forward rush and swallowed whole, is most frequently another fish, but any creature large enough to attract the pickerel's attention may fall prey to it.

Insect predators

Among the pond's most ferocious insect predators are the giant water bugs, of which there are a number of species, some over three inches in length. Giant water bugs grasp their prey—fish, frogs, tadpoles, other insects—with their powerful forelegs, paralyze them with injections of poison, and suck out their body juices.

A large diving beetle is a good example of the highly successful hunter, as is the equally large giant water bug. The latter is predacious and captures insects, tadpoles, fish, and even small frogs with its large, jackknife front legs. Once it holds its victim in its powerful grasp, the giant water bug injects a paralytic poison through its beak. If you pick one up, do so carefully, for the bug can give you a very painful bite.

The nymphs of the large dragonflies are among the most impressive underwater predators. Nymphs differ according to the kind of habitat in which they live. Stream and brook dragonfly nymphs are constructed quite differently from those of still waters. Depending on the species, a pond nymph may burrow into the mud, sprawl on top of the soft bottom, or cling to plant stems above the bottom. Most of these forms hunt primarily by sight, although some of the sprawlers depend more on touch.

Should you catch a large dragonfly nymph, place it in a container of pond water. If fed regularly on small worms or midge larvae, it should live successfully through the winter. It is well worth watching, for its feeding technique is amazing. The so-called underlip, or *labium,* of a dragonfly nymph consists of a long folded structure, hinged both where it joins the body and in the middle. At the tip is a pair of overlapping jaws.

Some nymphs stalk their prey, and others lie in wait. No matter which is the case, a sudden increase in blood pressure causes a lightning-fast extension of the labium when a victim comes within reach. The entire labium may extend an inch beyond the head of a very large nymph. The jaws at the tip of the labium, which are under the control of the nervous system, close on whatever they strike. The prey, usually another insect, a worm, or a crustacean, is then brought back to the mouth when a decrease in blood pressure allows the labium to fold under the head. While the true jaws and mouthparts of a nymph crush the food, the labium continues to hold it firmly in position. Further chewing goes on in a gizzard within the body of the insect.

If you keep nymphs in a shallow container of water and feed them enough, you should be prepared for the outcome of such feasting—a large nymph can expel a pellet of waste material at least a foot out of water.

146

Alert and able to strike with lightning speed, the dragonfly nymph is one of the pond's dominant insect carnivores. It preys on anything smaller than itself—other insects, crustaceans, snails, worms, and small fish. The nymph captures its prey with a specially adapted lower lip called a labium. Normally folded back under the head, the labium shoots far out to grasp the victim and drag it back to the mouth and jaws.

raw liver to a string and lower it to the bottom. When you pull it up later, it should be covered with flatworms if any are present.

Many turtles are active scavengers: snappers, musk turtles, and mud turtles are but a few that eat anything they come across. A great many bottom-dwelling immature insects crawl about, eating either plant detritus or larger plant remains and feeding upon animal carcasses that fall into the water.

All these relationships—phoresis, commensalism, mutualism, parasitism, predation, scavenging—are examples of major *ecological niches*, or ways of life. Every possible ecological niche tends to be occupied in a pond, and once again we are reminded that it truly is a world in miniature. Specializations for swimming, breathing, reproducing, and obtaining food are almost endless, yet they are easily observed by anyone who will take the time to look and to wonder.

These parasitic leeches, attached to the tail fin of a fish, represent one solution to the problems of survival faced by every pond creature. Parasitism is a specialized way of life and is most successful if the parasites do no great harm to the hosts, as is the case here.

Worlds within Worlds

On December 25, 1702, a remarkable Dutchman, Anton van Leeuwenhoek, wrote:

> Whenever I turned my attention to duckweed, I always noticed that it never grows in deep water, even though the water be small and stagnant, and without motion, save such as is imparted to it by the wind; but it is seen in great plenty on broad sheets of water, which are not deep and have little motion, but especially in narrow and shallow ditches. . . .
>
> I got some of the duckweed scooped out of this water in an earthen pot, with lots of water, so that their roots might not be hurt.
>
> I took several of these little weeds out of the pot of water with a needle, one after the other, as nicely as I was able to, and put them in a glass tube of a finger's breadth, that was filled to the top with water, and also in a smaller glass tube, and suffered their little roots to sink down gently; and then examining these roots with the microscope, I beheld with wonder many little animals of divers kinds, which escape our naked eye. . . .

Leeuwenhoek went on to describe several kinds of perit-rich protozoans, rotifers, and *Hydra*. He was the first to write

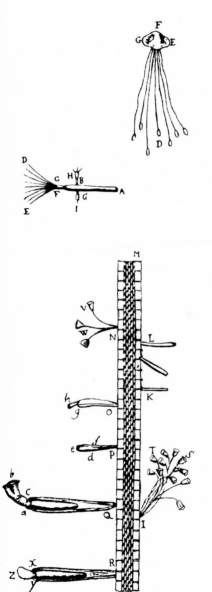

In 1702 Anton van Leeuwenhoek made the first known drawings of the duckweed microhabitat. Three are reproduced below (*from top to bottom*): a single duckweed plant with its trailing roots; a hydra with two buds almost ready to break away and become separate animals; a section of duckweed root with a number of living things growing on it, including peritrichs, tube-dwelling rotifers, protozoans, and diatoms.

about the fascinating miniature habitat, or *microhabitat*, created by the roots of duckweed.

Life under the duckweed

The microhabitat of a duckweed plant consists of one or more fine filamentous rootlets, less than half an inch long, hanging down under the water's surface film, shaded and protected by the small leaves floating overhead. Here an organism is exposed to plenty of diffused light and plenty of oxygen, but it is not in great danger of being eaten unless a passing duck scoops up the entire duckweed plant.

The plants and animals of the duckweed microhabitat are almost all attaching forms. Desmids, diatoms, and filamentous blue-green algae often are stuck to, or wound around, the roots. A large number of peritrichs, with their bell-shaped ciliated mouths, are always present. Some, such as *Vorticella*, are found singly; some form colonies. Moving up and down the roots are other protozoans of many kinds, which feed on bacteria, algae, and tiny fragments of detritus. Tube-dwelling rotifers and fixed, crowned rotifers extend out from the roots. There are also rotifers that move about. Then there is *Hydra*, the graceful coelenterate with long, drooping tentacles held out in a delicate net to snare passing copepods and water fleas. Predacious insect larvae wriggle from one cluster of roots to another, picking away at the animals attached to the plants. The small leaves overhead and the even tinier watermeal plants may be packed so close together that they form a dry, unbroken canopy on top of which run insects and spiders. Watermeal would not grow so well in some ponds without duckweed, for the larger plants keep the smaller ones from being blown ashore.

So it is evident that where there is duckweed there will also be a host of other plants and animals living close to the surface of the pond. In a cupful of water, you can easily bring home for study a bit of this miniature pond world.

In the miniature jungle of the duckweed community, stalked, bell-shaped *Vorticella* crowd the lower surfaces of the matchhead-sized leaves, and an insect larva thrusts its way between the roots. A *Hydra* (*below*) spreads its sting-laden tentacles among glasslike strands of the alga *Spirogyra*, and funnel-shaped protozoans called *Stentor* swarm through the water.

162

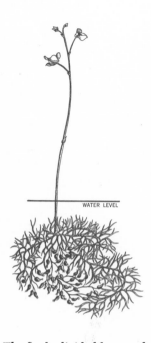

WATER LEVEL

A deathtrap for a home

Bladderwort is another plant that creates a microhabitat for plants and animals. Tangled masses of its stems provide living space for countless fixed, or *sessile*, organisms. Indeed, more sessile rotifers—both numbers of species and numbers of individuals—attach to bladderwort than to any other aquatic plant. On the other hand, waters immediately surrounding bladderwort contain only one-tenth as much plankton as waters farther away. Bladderwort has bladders, or traps, with spines and hairs arranged in definite patterns around their mouths. Small organisms either are caught in the spaces between the hairs or are actually drawn into the traps. A bladder is normally compressed; but when a small creature touches one of its hairs, it suddenly inflates, sucking into its opening whatever is outside. Once inside a bladder, the trapped creature cannot escape, because a transparent valve guards the opening. It dies and is digested and then absorbed by special hairs projecting into the bladder, and so it becomes a part of the plant's food.

The finely divided leaves of bladderwort (*opposite page*) are studded with rows of deathtrap bladders, each about a tenth of an inch across. The overall plant (*above*) consists of free-floating, branching stems up to three feet long. Aerial stalks bear half-inch flowers which may be blue, yellow, purple, or white, according to species. Each bladderwort bladder is a flattened sac which suddenly expands to suck in any small swimming organism that disturbs the sensitive hairs around its opening. The bladder resets and is ready for more captures in about twenty minutes. The single bladder shown in the close-up (*right*) serves as a base for two tube-dwelling rotifers.

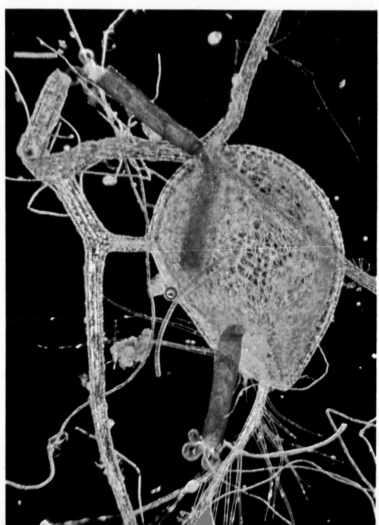

A stand of white water lilies forms a world within the world of the pond. The plants' pads provide coves of quiet water in which duckweed and watermeal grow; their water-repellent upper surfaces serve as islands of dry land for insects ranging from dragonflies, like this one at the right of the turtle (*above*), to pinhead-sized jumping plantlice (*below*).

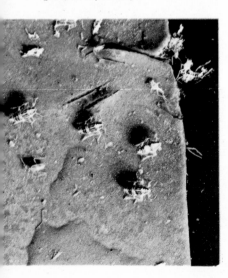

A *floating home*

You should not have much difficulty finding a good growth of white pond lily, floating heart, or water shield. A large floating leaf serves as a microhabitat for a surprising number of small animals and plants.

From a few feet away, a lily pad does not seem to be the home of anything more than a few insects. You can see dragonflies and damselflies resting on it briefly and watching for flying prey. As you draw nearer, you may notice a green bug or a ladybird beetle walking about. Perhaps the surface of the leaf is speckled with small white dots; as you reach out to bring the leaf closer, all the dots bound into the air and land on nearby floating leaves or bounce and hop across the surface of the water. You need a magnifying glass to see that they are tiny insects, jumping plantlice and springtails.

A jumping plantlouse uses its legs to leap into the air, but a springtail has a unique apparatus for literally catapulting itself through the air. Two long filaments projecting from the end of the springtail's abdomen are tucked beneath a knob on the underside of its abdomen. When certain muscles contract, the filaments grow tense, but they do not slip off the knob until enough force has been built up for them to snap off violently. When that happens, the springtail shoots off

166

packed eggs that drift over the surface. The minute larvae hatch through an opening in the bottom of the eggs and emerge directly into the water. Certain midges anchor submerged egg masses with filaments attached to water-repellent disks that float at the surface. Hanging underneath their buoys, the eggs are carried about by winds and water currents. Females of other midges belonging to the same family fly over the water and swish the tips of their abdomens in the water as eggs are released.

The depths of the pond

In the deepest portions of the ocean there are living things unlike any seen around the seashore. Though ponds are not very deep, a somewhat similar situation exists in their benthic, or bottom, regions. This situation is due partly to a lack of light. If a pond is productive, there is so much phytoplankton and zooplankton suspended in the water that only a small amount of light reaches the bottom. This means

Dragonflies also engage in mating flights with the female clasped behind the head by the male. Here, the pair has alighted for a moment, and the female (on the left), to fertilize her eggs, has brought the tip of her abdomen around to a special pocket beneath the male's abdomen in which he has previously stored capsules of sperm. Unlike damselflies, many species of dragonflies simply scatter their eggs at random in the water.

177

that on the bottom very little photosynthesis goes on, and as a result the oxygen content of the water is low. Even so, animals living there have some advantages. Because of the dim light a vulnerable bottom dweller cannot be seen easily, and it has every opportunity to burrow and hide. It does not have to use much energy in moving about; it can spend most of its time resting quietly or walking slowly. Food is plentiful: the bottom is rich in fine detritus, and larger organic particles and dead animals continually rain down from above. For predators, the bottom mud abounds with small animals—large numbers of individuals, if not of species.

Ponds generally are basin-shaped. Fine sediments constantly settle along the sloping sides and in the center. Thick, rich blankets of organic sediment do not make the best of habitats for most pond organisms; in fact, many pond animals suffocate in such surroundings. There are a few organisms, however, that are adapted to this kind of habitat and are not found elsewhere in great numbers.

The burrowers of the bottom

Deep in the soil on land you would expect to find roundworms, or nematodes, and segmented worms. In a pond bottom, too, microscopic nematodes are often present in enormous numbers, and there may be almost as many segmented worms. Both types of worms thrive in regions that are thick with sediment and that have diminished supplies of oxygen. The nematodes are burrowers that probe their way through the mud with needlelike bodies. They wriggle slowly back and forth as they go along but increase their speed if they come to a region where the particles of sediment are not tightly packed.

Some fresh-water segmented worms, or oligochaetes, make burrows with cylindrical extensions rising above the bottom. They live and feed head downward in the burrow, passing quantities of mud through their digestive tracts and finally excreting it from a pore at the end of their bodies, which stick far out of the tubes. From the mud they extract nourishing organic matter. A bottom infested with worms of this sort—*Tubifex* is a common type—is a striking sight, for thousands of long swaying shapes writhe about, their movements causing some exchange of water in their burrows.

Other kinds of worms burrow freely through the bottom, feeding on detritus as they go along. One variety, *Chaetogaster*, is carnivorous and feeds upon smaller worms, small crustaceans, and burrowing insect larvae, mostly midges.

More bottom dwellers

There are also other burrowers, including mussels, snails, copepods, nymphs of some insects, and midge and horse-fly larvae. In addition to animals that burrow into the bottom, there are creatures that sprawl on top of it—certain mayfly nymphs, dragonfly nymphs, crayfish, isopods, snails, and flatworms.

A few larger animals get most of their food from the bottom. Snapping and musk turtles capture and scavenge what they can. Suckers, catfish, and carp are well adapted to rooting about and seek food either with sensitive *barbels*, fleshy sensory organs that hang like whiskers under their mouths, or with mouths that are directed straight down.

By far the greatest activity in the bottom goes unseen: that of bacteria, fungi, and protozoans, which live successfully with little or no oxygen. Their activities are of inesti-

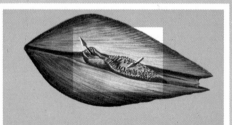

Here is a close-up view of a fresh-water clam's siphons. Water enters through one fringed, slitlike opening, or siphon, and leaves by another smaller opening just above it. Suspended food particles are captured on the mucus-coated gills (the orange areas within the siphons). The same water currents also bring the clam oxygen and carry away its waste products.

mable importance to the biological balance of a pond, for they alter much of the organic matter that falls to the bottom and release nutrients into the water that otherwise would remain locked in the sediment.

Animal strainers

Some of the animals of a muddy bottom, and many of those of a cleaner bottom, are filter feeders: they draw in water from which they extract food particles. Perhaps the most striking example of a filter feeder is the fresh-water mussel. This mollusk, which actually is a true clam, pulls itself along through the sediment by extending a blade-shaped muscular foot that swells at the tip and then contracts. At the opposite end of its shell, elevated above the bottom, two fleshy openings, or *siphons*, lead to the interior. If you look through a magnifying glass at a mussel in an aquarium, you will notice that water flows into the larger, fringed lower siphon, and out of the smaller, slightly longer siphon.

The inner surface of the fleshy mantle that surrounds the mussel's body, the surface of the body itself, and the gills are covered with cilia similar to those of ciliate protozoans. The cilia beat in such a way as to force water through pores in the gills. As the water flows through, mucus secreted by the gills traps particles of detritus, zooplankton, and phytoplankton in thick, sticky sheets. Other cilia on both sides of the animal pass the sheets of mucus toward the mouth, which lies between two leaflike flaps on each side. The filtered water is exhaled through the upper siphon. The quantities of plankton and suspended detritus removed from a pond by fresh-water mussels are truly enormous, especially when four or five of these mollusks are crowded together in one square foot of pond bottom.

The far smaller pill clams, pea clams, and fingernail clams have feeding habits similar to those of the fresh-water mussels, and they can also filter tremendous amounts of detritus. Occasionally as many as five hundred individuals may be packed in a square foot of pond bottom. You can hold dozens of these little clams in the palm of your hand, but it is best not to try to keep them, for they usually die rather quickly in home aquariums. Mussels live quite well in aquariums, although it is almost impossible to feed them adequately; and you must immediately remove those that die, before they decay and pollute the water.

181

Stentor, shown here fifteen times life size, are trumpet-shaped protozoans which are found both as free-swimming individuals and, as here, clustered together in a crowded colony on a bit of pond substrate. The wide end of the animal is fringed with a circle of cilia which set up currents to sweep food organisms into the mouth opening. The bright green coloration of most of the *Stentor* shown here results from the presence of symbiotic algae living within the animals' bodies.

There are multitudes of smaller filter feeders on every exposed surface in a pond. All the peritrich protozoans, which may grow singly or in colonies, strain and tug at their filamentous attachments as they pull in suspended particles with their strong ciliary currents. Sometimes there are so many individuals of the larger protozoan *Stentor* that they cover sticks and pebbles with a thick coating. Tube-dwelling rotifers also create ciliary whirlpools that bring water-borne matter their way. Bryozoan "moss animals," such as the massive *Pectinatella* colonies, strain quantities of water wherever they grow. Sponges are among the most effective animal strainers, with their thousands of pores and flagellated chambers.

There are many other kinds of filter feeders in a pond, but none are more active than certain members of the zooplankton. A particular group of copepods (calanoids), ostracods, and cladoceran water fleas feed only by sifting organisms and material from the water. The feeding appendages of some of these small crustaceans have hairs so fine and so closely spaced that they effectively filter out individual bacteria. Calanoids use a highly complex system of antennae and other appendages to bring a stream of water into their filtering apparatus at a rate of over one thousand beats a minute.

with sicklelike jaws. This is the larva of a tiger beetle, which is a common resident in the wet sand of pond shores. It rests with its head level with the surface. When another insect runs by, the larva reaches out and grabs it. It maintains a secure anchorage in its hole with the bumps, or tubercles, on its back; these also enable it to go up and down its burrow quickly.

Dig a hole in the wet sand not far from the water's edge; it will quickly fill with water seeping through the sand. If you take some of this water and strain it through a fine cloth, you should collect a number of organisms that live in the water between sand grains. These animals usually are elongated and able to slip through narrow spaces with ease. They are roundworms, segmented worms, rotifers, gastrotrichs, water bears, and highly specialized harpacticoid copepods.

Here is a victim's-eye view of the tiger beetle larva's head. The scraggly covering of white hairs helps protect the creature from the direct rays of the sun. The adult beetle also has long sicklelike jaws and is just as predatory as the larva.

Bacteria, concentrated mostly in the upper inch of sand, are the most numerous organisms. Many millions of them may inhabit a cubic half inch of wet sand. In the same volume, there can also be several thousand protozoans, a dozen or so rotifers, and a few copepods and other organisms. Altogether the sand grains shelter a rather crowded community but a flourishing one.

Exploring the pond world

We are, by nature, explorers, but not all of us can go to faraway lands, to the depths of the oceans, or into space. The odd thing is that man has by-passed so many obvious places in his search for the new. And no one realizes this fact more than the biologist with an interest in fresh-water life. He will tell you that in even the smallest pond you can find many things that presently cannot be explained. Each year professional journals are filled with new studies of fresh-water organisms. Some pond animals have not even been named by biologists, much less studied.

The organisms described here are but a very few of those that may be found in any pond in your neighborhood. A pond is accessible to anyone, and knowledge of its inhabitants can be acquired with nothing but a pair of eyes and an interested, inquisitive mind. Nets, magnifying instruments, and glassware are all luxuries. A source of information about what you observe is essential, however. Books open the door to the pond world, and you will find many that lead you far beyond the brief introduction presented here.

Good luck in your explorations. A final word of warning: once you become a pond hunter, you may remain one all your life.

Beneath its surface, a pond harbors a miniature world
which in many ways is a model of the larger world
beyond its shores. To anyone who will take the
time to stop and look into its waters, a pond offers
a vivid panorama of the living world.

Appendix

Ponds in Our National Parks and Wildlife Refuges

Spreading across America is a splendid system of national parks, monuments, and wildlife refuges, maintained and operated by the Department of the Interior. The parks and monuments, administered by the department's National Park Service, preserve sites of special scenic or historical interest and provide facilities for human recreation and enjoyment. In the wildlife refuges, administered by the department's Fish and Wildlife Service, human interests come second: these areas are intended primarily for the shelter and protection of our native birds and animals. Throughout the system, you will find ponds of every description, in natural, unspoiled settings.

The majority of these ponds are very much like the ones described in this book, since ponds exhibit a remarkable constancy throughout the country. There are exceptions, however, and fascinating ones—the thermal pools of Yellowstone National Park, where the water remains close to the boiling point; the brine-filled hollows of Death Valley National Monument; the frigid tundra ponds of Mount McKinley National Park. All these support living communities adapted to their unusual conditions.

So many parks and refuges contain ponds that this book could not describe them all, but those discussed here will provide a representative sampling of the many such areas administered by the Department of the Interior. It is hoped that you will take the opportunity to visit some of them and to enjoy their ponds.

Aransas National Wildlife Refuge (Texas)

This wildlife refuge, fringing the Gulf coast of Texas, is perhaps the most famous of all, since it is the winter home of the most publicized and one of the least populous of all native American birds, the whooping cranes. In addition to the celebrated whoopers, a rich variety of other wading birds pass the winter in the refuge's seven ponds. The refuge is open to visitors throughout the year, although the area frequented by the forty-odd whooping cranes still in existence is strictly off-limits.

Bitter Lake National Wildlife Refuge (New Mexico)

Bitter Lake includes some eighteen pothole ponds formed by an unusual process. Underground water dissolved gypsum layers in the soil, and the domes of the resulting hollows eventually

WHOOPING CRANE

202

caved in to form ponds varying in depth from fifteen to nearly one hundred feet. The high salinity of the water in these ponds supports the growth of marine algae, over five hundred miles from the sea. Bitter Lake is a happy hunting ground for bird watchers: on nearly any day during the winter months, from fifty to sixty species can be observed within the refuge.

Cape Hatteras National Seashore (North Carolina)

In wet weather, more than a hundred ponds are scattered through this seventy-mile chain of barrier islands extending along the Carolina coast; in dry weather, the number shrinks to a few dozen. Periodic invasions of salt water into normally fresh-water ponds are common. Despite the changeable conditions, or perhaps because of them, plant and animal life is both richly varied and abundant. Although Hatteras is separated from the mainland, the presence of the salamander *Amphiuma,* which cannot tolerate salt water, suggests that a connection once existed. Cape Hatteras National Seashore includes Pea Island National Wildlife Refuge.

Carolina Sandhills National Wildlife Refuge (South Carolina)

This refuge contains twenty-seven fresh-water ponds, all manmade, varying in size from one to sixty acres and ranging in depth from less than a foot to eighteen feet. The pond life is highly typical of ponds everywhere, and a large percentage of the plants and animals pictured in this book can be observed here. A feature of the refuge is a self-guiding nature trail along which visitors can enjoy many of the refuge's attractions. With written permission from the refuge office, hikers may visit other parts of the refuge's 46,000 acres.

BLACK DUCK

Clarence Rhode National Wildlife Range (Alaska)

How many ponds here? It is unlikely that anyone will ever count them to find out: estimates suggest that the range's 1,800,000 acres include over a million ponds, ranging from miniatures only a few yards across to shallow lakes fifteen miles long. The ponds are summer breeding grounds for tens of thousands of geese and ducks and numerous species of other waterfowl.

Columbia National Wildlife Refuge (Washington)

The Columbia Refuge includes over one hundred ponds, which collectively illustrate how ponds that are more or less side by side can vary radically in their animal and plant populations. They host a variety of wildlife, including muskrats, beavers, raccoons, skunks, badgers, and over 160 species of birds. Some of Columbia's ponds teem with fish, and others contain no fish at all. Why? The degree of acidity varies from pond to pond;

203

perhaps this is sufficient to make subtle differences in food webs that encourage fish proliferation in some ponds and discourage it in others.

Everglades National Park (Florida)

Located at the very tip of the Florida peninsula, the Everglades offers exotic wildlife that cannot be found anywhere else in the country. Among the common reptiles are alligators; snakes; and red-bellied, snapping, and stinkpot, or soft-shelled, turtles. The park includes countless ponds, some of them created by fire—hollows formed from the burning of peat—and some produced by alligator digging. Water life of every sort is abundant here, although its concentration in ponds varies with season.

ALLIGATOR

Great Swamp National Wildlife Refuge (New Jersey)

This wildlife refuge, just forty miles due west of Times Square, is an important stopover point for migratory waterfowl. Six ponds are included within its bounds, which will eventually embrace over five thousand acres of unspoiled wilderness located in the center of one of the most densely populated areas in the country. In addition to birds, reptiles and amphibians are plentiful here: eight species of turtles and twenty-one of frogs, toads, and salamanders can be found in Great Swamp.

Gulf Island National Wildlife Refuge (Mississippi)

Located approximately ten miles from the mainland, this refuge can be reached only by boat. It includes about forty ponds, the waters of which range from slightly brackish to a saltiness near that of seawater. As a result, the plant and animal populations are an odd mixture: familiar fresh-water forms such as cattails and raccoons share living quarters with such marine forms as black drum and flounder. The islands are closed to the public during the fall and winter months, to minimize disturbance to the great flocks of water birds that nest there during that part of the year.

Lacreek National Wildlife Refuge (South Dakota)

Lacreek contains twelve fresh-water impoundments designed specifically to accommodate migratory waterfowl, but housing a wide variety of familiar pond animals as well, including beavers, raccoons, snapping turtles, and bullfrogs. Common plants include sedges, cattails, bulrushes, spike rush, and pondweed. Over two hundred species of birds have been identified within the refuge. An attempt is being made to establish a nesting colony of the once nearly extinct trumpeter swan. The impoundments can be visited at all times of the year, but visitors must first obtain permits from the refuge manager.

204

Olympic National Park (Washington)

Olympic National Park includes dozens of ponds, mostly of glacial origin and ranging in elevation from near sea level to over a mile above. Many of the ponds at higher altitudes display striking color contrasts owing to vegetation and silt deposits. Among the park's prime animal attractions is the Roosevelt elk, which is often found in numbers around the higher ponds.

Savannah National Wildlife Refuge (South Carolina)

The thirteen ponds included in this refuge, ranging in size from twenty-five acres to nearly five hundred, are connected to the tidal portion of the Savannah River, and thus have daily tides of from six to nine feet. No less than 212 species of birds, including many waterfowl, have been identified within the refuge. During the summer white ibises and purple gallinules are commonly seen, and the refuge is an important resting and feeding place for thousands of ducks. No hunting is permitted here, but the ponds are open to fishermen from March 15 to October 25, and offer bluegill, largemouth bass, bowfin, catfish, carp, crappie, and gar.

BLACK CRAPPIE

Sherburne National Wildlife Refuge (Minnesota)

One of the newest of our refuges, Sherburne National Wildlife Refuge contains about a thousand ponds, ranging in type from prairie potholes to typical woodland ponds. Many are the remains of what were once bog lakes. Common birds include ducks, herons, bitterns, and the short-billed marsh wren.

Yellowstone National Park (Wyoming, Montana, Idaho)

The oldest and still the largest of our national parks, Yellowstone is best known for its geysers, hot springs, and bears, but it also contains several hundred ponds of all sizes and types. Many of these ponds are the result of beaver activity, and Yellowstone is a fine place to observe these interesting mammals in action. Muskrats, moose, many kinds of ducks, Canada geese, and trumpeter swans can also be seen. Even the hot springs are not entirely devoid of life: though the water is near the boiling point, certain primitive algae thrive in them.

Yosemite National Park (California)

The wild beauty for which Yosemite National Park is noted is the result of glacial action, and among the work of the glaciers are several dozen ponds. Some are quite shallow, and others are deep enough to show temperature layering of the water, an effect more common in larger bodies of water. Frogs, toads, and salamanders are commonly found. Yosemite is a hiker's paradise with trails leading up to many scenic alpine ponds.

205

What a Fish Sees

What does a fish see when it surveys its underwater world? No one can say for certain, of course, since one cannot get inside a fish's brain, but it is possible to make some deductions based on laboratory studies of fish behavior and on what we know about the nature of light.

First, we know from laboratory experiments that fish see in color, although not necessarily in quite the same way that humans do. The black bass, for example, sees colors about as you would if you were wearing yellow sunglasses. Fish are probably rather near-sighted, seeing objects at close range more clearly than those farther away. And some species, at least, apparently have well-developed depth perception.

A fish can see objects on the shore, too. Why is it that you seldom see a fish from the bank of a pond? Except for those sunfish and bass that are guarding nests close to shore, all you see is a cloud of mud rising from the bottom where a fish, with a sudden burst of speed, darts away to deeper water. The explanation lies in the fact that a fish usually sees you long before you are aware of it. But how can a fish see you if it lies below an overhanging bank, not in your direct line of sight? To understand that it does see you in this position, you need to know something of elementary optics. Light passes through a transparent medium, such as air, water, or glass, at a speed governed by the composition and density of the medium. Water is more dense than air, and light is slowed and deflected when it enters water. This accounts for the apparent bending of a stick when you place it at an angle half in the water and half in air. The greater the angle, the greater the apparent bending. There is, however, a limit to the angle of refraction, beyond which light, instead of being refracted, is reflected—bounced off the underside of the water surface, as though it were the surface of a mirror. This critical angle is 48.8 degrees.

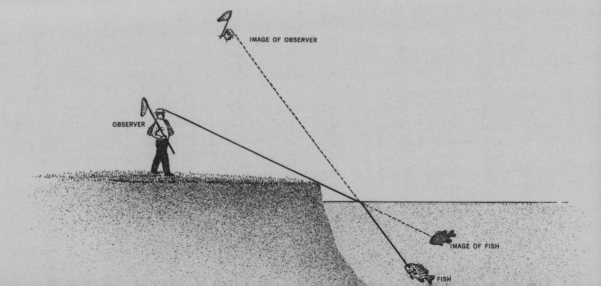

Thus, a fish looking up toward the surface in quiet water sees the shoreline through a sort of transparent window, 97.6 degrees wide. Beyond this window he sees mirror images of objects (such as the rock in the diagram) within the pond. The secret, then, of approaching such a fish is to keep yourself out of the "window" portion of its visual field. In this way, the fish will not be able to see you, though of course you will not be able to see the fish, either. At any rate, this is a useful way of getting to a likely fishing spot without frightening away too many fish.

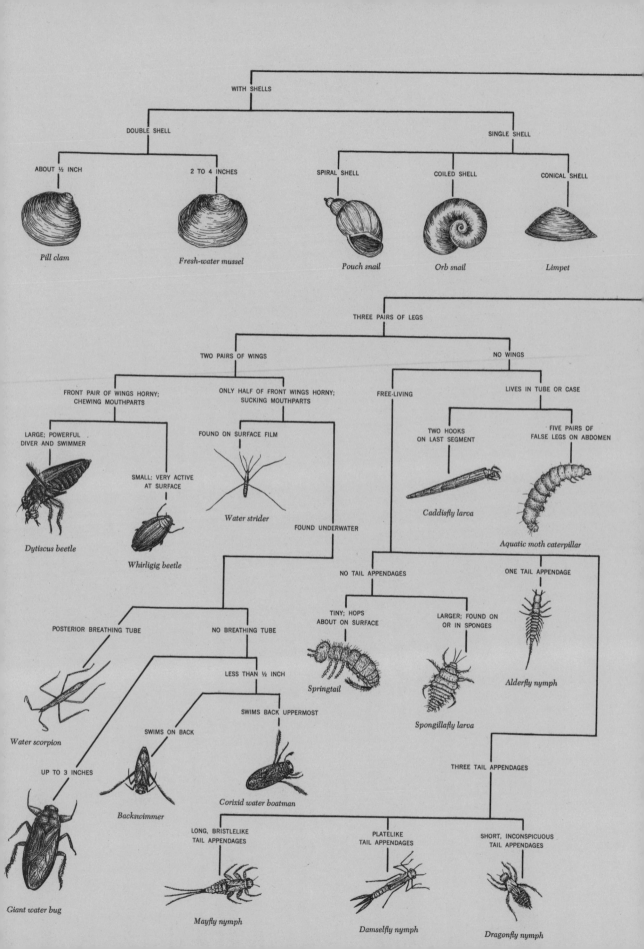

WITH SHELLS

DOUBLE SHELL

SINGLE SHELL

ABOUT ½ INCH

2 TO 4 INCHES

SPIRAL SHELL

COILED SHELL

CONICAL SHELL

Pill clam

Fresh-water mussel

Pouch snail

Orb snail

Limpet

THREE PAIRS OF LEGS

TWO PAIRS OF WINGS

NO WINGS

FRONT PAIR OF WINGS HORNY;
CHEWING MOUTHPARTS

ONLY HALF OF FRONT WINGS HORNY;
SUCKING MOUTHPARTS

FREE-LIVING

LIVES IN TUBE OR CASE

LARGE; POWERFUL
DIVER AND SWIMMER

FOUND ON SURFACE FILM

TWO HOOKS
ON LAST SEGMENT

FIVE PAIRS OF
FALSE LEGS ON ABDOMEN

SMALL: VERY ACTIVE
AT SURFACE

Water strider

Caddisfly larva

FOUND UNDERWATER

Dytiscus beetle

Whirligig beetle

Aquatic moth caterpillar

NO TAIL APPENDAGES

ONE TAIL APPENDAGE

POSTERIOR BREATHING TUBE

NO BREATHING TUBE

TINY; HOPS
ABOUT ON SURFACE

LARGER; FOUND ON
OR IN SPONGES

LESS THAN ½ INCH

Springtail

Alderfly nymph

SWIMS BACK UPPERMOST

Water scorpion

SWIMS ON BACK

Spongillafly larva

UP TO 3 INCHES

Corixid water boatman

THREE TAIL APPENDAGES

Backswimmer

LONG, BRISTLELIKE
TAIL APPENDAGES

PLATELIKE
TAIL APPENDAGES

SHORT, INCONSPICUOUS
TAIL APPENDAGES

Giant water bug

Mayfly nymph

Damselfly nymph

Dragonfly nymph

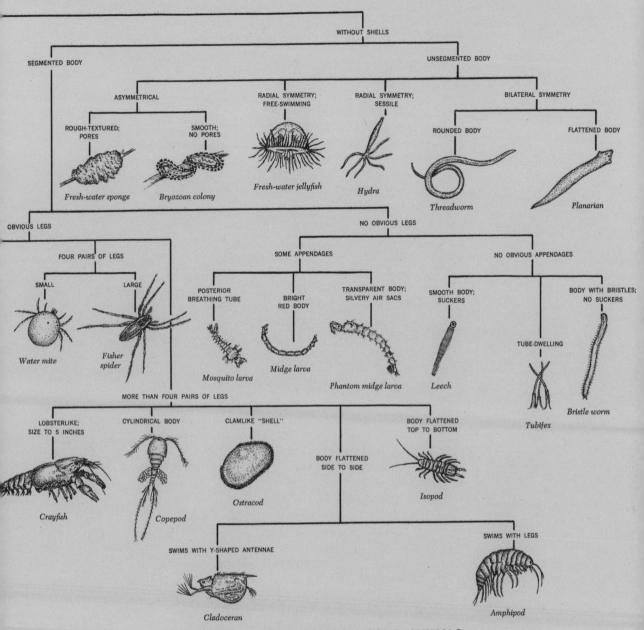

WITHOUT SHELLS

SEGMENTED BODY — UNSEGMENTED BODY

ASYMMETRICAL — RADIAL SYMMETRY; FREE-SWIMMING — RADIAL SYMMETRY; SESSILE — BILATERAL SYMMETRY

ROUGH-TEXTURED; PORES — SMOOTH; NO PORES — ROUNDED BODY — FLATTENED BODY

Fresh-water sponge — *Bryozoan colony* — *Fresh-water jellyfish* — *Hydra* — *Threadworm* — *Planarian*

OBVIOUS LEGS — NO OBVIOUS LEGS

FOUR PAIRS OF LEGS — SOME APPENDAGES — NO OBVIOUS APPENDAGES

SMALL — LARGE — POSTERIOR BREATHING TUBE — BRIGHT RED BODY — TRANSPARENT BODY; SILVERY AIR SACS — SMOOTH BODY; SUCKERS — BODY WITH BRISTLES; NO SUCKERS

Water mite — *Fisher spider* — *Mosquito larva* — *Midge larva* — *Phantom midge larva* — *Leech* — TUBE-DWELLING

MORE THAN FOUR PAIRS OF LEGS

Tubifex — *Bristle worm*

LOBSTERLIKE; SIZE TO 5 INCHES — CYLINDRICAL BODY — CLAMLIKE "SHELL" — BODY FLATTENED TOP TO BOTTOM

BODY FLATTENED SIDE TO SIDE

Crayfish — *Copepod* — *Ostracod* — *Isopod*

SWIMS WITH Y-SHAPED ANTENNAE — SWIMS WITH LEGS

Cladoceran — *Amphipod*

A GUIDE TO SOME COMMON POND ANIMALS

The pictorial key presented on these two pages will help you identify some of the more common creatures you may encounter in your pond explorations. By noting the physical characteristics of the unknown animal and using these characteristics to trace a path down the branching forks of the key, you will eventually reach a "dead end" that identifies the animal. The key is limited to non-microscopic invertebrates and shows only typical examples of the various animal groups included. More elaborate and comprehensive keys to aquatic plants and animals can be found in many of the books listed in the Bibliography on page 227.

How to Learn More About a Pond

If the topics discussed in this book have interested you, the best way to further that interest is to get out and study ponds yourself. The most fruitful way to go about this is to select one pond to which you have easy access and to learn everything you can about it. By finding out how a single pond changes from season to season, and from year to year, you will build up a very good understanding of the natural history of ponds in general.

Accurate, permanent records of your observations and measurements are of utmost importance. For a start, you will want a map of your pond; you can sketch out one yourself, or you may find that your County Clerk's office can tell you where to obtain a printed map that includes your pond. If you have the use of a boat, you can plot the contour of the pond's bottom by making depth measurements with a pole or weighted line marked off in six-inch intervals.

A camera, even of the simplest type, is a useful tool in pond study. By periodically taking pictures at a series of fixed points around the pond margin, you can build up a valuable photographic record of seasonal vegetation changes and measure the rate of yearly succession. Numbered stakes, driven firmly into the ground, can serve as permanent points of reference for your picture taking.

Certain physical data can be added to your volume of information about your pond. A regular record of water temperature is useful. A simple pH test kit, sold by tropical fish dealers, will let you measure the acidity or alkalinity. A gardener's soil-testing outfit can be used to measure other chemical aspects of the water. You can measure turbidity by lowering a metal disk painted white (the top of a large tin can) and noting the depth at which it disappears from view.

Much of your attention will of course be directed toward the plants and animals that make up your pond's living community. At the least, you will want to make a census of the flora and fauna of the pond, and perhaps to keep records of the fluctuations in the abundance of various species. As a more ambitious project, you might try working out a food-web diagram for your pond. Or you may decide to concentrate on one segment of the population—the dragonflies, for example—and learn all you can about it.

The equipment you will need for your pond explorations is neither complicated nor expensive, and much of it can be homemade. Some of these devices are described on the following pages.

Basic collecting equipment

One of the most useful items for pond study is an ordinary white enamel or plastic pan. To use it, simply dump a handful or so of dead leaves and other debris from the pond bottom into the pan and then pick the pieces out one by one, examining each for eggs, animals, and other items of interest. A folding pocket magnifying glass is helpful. Specimens can be transferred from the pan to jars or vials with a pair of forceps or a medicine dropper. A plastic basting gadget, shaped like a giant medicine dropper and available in most variety stores, is handy for picking up hard-to-catch creatures such as leeches and some insect larvae.

Weed grapple

With this device, fashioned from a short length of pipe and some stiff wire, you can bring up samples of submergent vegetation and then place them in your collecting pan for detailed examination.

Sorting screen

This device is especially useful for collecting burrowers and bottom dwellers. It consists of a simple wooden frame of any convenient size, with ordinary window screening tacked to the bottom. Place a quantity of bottom debris on the screen and slosh water over it to wash away the mud and silt.

Nets

The most useful net for pond study is a long-handled dip net, either purchased or homemade, sturdy enough to withstand a fair amount of hard use. Swept through plant tangles and along the bottom, the dip net may yield fish, amphibians, crayfish, insect larvae, and the like. An ordinary kitchen strainer, attached securely to a broom handle, is a good substitute. A six-foot minnow seine, operated by two people, allows a more wholesale sampling of the pond's fish population, but do not handle or remove fish needlessly. You can identify and count them, without harming any, simply by drawing the net in so they have only a few inches of water in which to swim; then release them by raising the net. If you are using a microscope in your pond explorations, a plankton net made of fine-mesh nylon or muslin will be useful for concentrating the microscopic plants and animals of the pond's limnetic zone.

Artificial substrates

One method of collecting sessile organisms such as sponges and bryozoans is to secure pieces of artificial substrate in various parts of the pond and simply leave them for a period of weeks or months. Pieces of wood, slate, asbestos house siding, or similar materials can be used for this purpose.

Glass-bottomed box

To observe life underwater, you will need some means of eliminating the glare and distortion caused by the water's surface. A glass-bottomed box answers this need, and can be easily made from any watertight wooden box. Cut a window in the bottom of the box and fit it with a piece of glass held in place with strips of wood and calked with waterproof putty or aquarium cement. Painting the inside of the box black will reduce glare and make the device more effective. Try using the box at night as well as during the daylight hours, with a waterproof flashlight (or an ordinary flashlight in a glass jar) as a light source.

212

Temperature-measuring equipment

Here is an easy-to-make device that will allow you to measure the water temperature at any desired depth. Simply lower the bottle to the required depth, pull out the stopper, and, when the bottle has filled, haul it up and read the temperature on the thermometer.

Bottom sampler

A simple dredge for obtaining samples of the pond's bottom material can be made by bolting a No. 10 can—the size used for fruit juices—to a broom handle or other pole. Flatten the mouth of the can somewhat, and punch holes in its bottom so excess water can drain off. Use the device with the same sort of motion you would use for raking leaves. Dump the bottom sample into your collecting pan or sorting screen for further processing.

Turtle traps

Turtles are among the wariest of pond animals, but you can sometimes capture them by nailing a bushel basket or similar container to the side of a favorite basking log. Your sudden appearance on the side of the pond away from the trap may startle one of the shy animals into tumbling into the basket.

Homemade Ponds

There are so many joys in having a small body of water for your very own, and so much to learn from it, that you should think seriously about the possibilities of creating one. A birdbath in a city yard is a pond and usually a good one. Birds making use of it bring cysts, spores, seeds, eggs, and microorganisms on their feet from other, distant bodies of water. Soon the water is swarming with a teeming plankton community in miniature. Even a large plastic washbasin left out and kept filled will after a while contain organisms that arrive by means of airborne spores and cysts. Flying aquatic insects—beetles, water boatmen, midges, and many others—may be attracted to the water's reflective surface.

As a slightly more ambitious project, you might consider creating a pond by excavating a pit in your backyard and placing in it a somewhat larger container—a plastic wading pool, for example—flush with the ground. If possible, select a site that gets some direct sunshine and some shade during the course of the day: too much sun can lead to overheating problems and algae population explosions.

This type of created pond lends itself well to study and observation for two reasons. First, you have more or less complete control over its contents and physical makeup, something you could not possibly achieve with a full-sized pond. Second, and more important, you have the advantage of being able to watch its ecological succession right from the very beginning, and a fascinating story will unfold as your pond develops and gradually takes on its own unique personality.

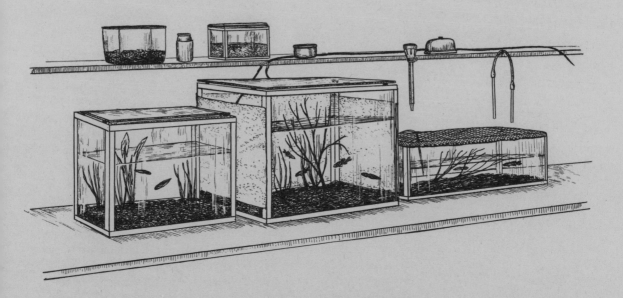

Indoor artificial ponds offer interesting projects, too. An ordinary aquarium—or better still, several aquariums—filled with pond water, a little mud and detritus from the bottom, a few selected small animals from your local pond, and some submergent plants can provide a source of year-round interest. If a sheet of glass is kept on top of the aquarium, with only a slight airspace, evaporation will be kept to a minimum and flying forms cannot escape. As you watch this miniature world month after month, you will notice patterns and cycles in the appearance of plant and animal forms.

The effectiveness of your indoor ponds will be enhanced if you observe a few simple rules. Do not add too much mud and detritus: roughly eighty percent of the aquarium's contents should be water. Direct sunlight is not necessary for the aquarium; the light from a north window is ideal. If the water becomes green too quickly with algae, cut down the light by masking the side of the aquarium toward the window with paint or paper. Most pond organisms thrive best in fairly cool waters, seventy degrees or less. Aeration is helpful but not necessary, and can be provided by an inexpensive diaphragm-type air pump of the sort used by tropical fish fanciers. Do not crowd too many animals into one tank or mix species that are not compatible. Other containers besides aquariums can be pressed into service; jars and jugs of all sizes can be set up as pond habitats.

Some naturalists have maintained such miniature ponds for years. They are far more instructive than a crystal-clear tank filled with gaudy tropical fish.

215

Keeping Pond Animals in the Home

A wide variety of pond animals, both large and small, can be maintained in captivity with a minimum of care and equipment. Keeping such creatures in your home, under conditions approximating their natural habitat, will give you the opportunity of studying them at close range and in detail. Given adequate care and feeding, many will carry out their complete life histories in captivity.

Before the specific needs of a few representative animals are discussed, some points of what might be called "pond watcher's etiquette" should be mentioned. First, you should never bring home more animals than you can properly care for. If you decide to maintain an animal in captivity, learn all you can about its needs and satisfy them to the best of your ability. If you cannot do so, or if your curiosity about the animal has been satisfied, release it in the same place you captured it. Occasionally you may encounter an individual animal that, for no observable reason, does not respond to home care—a turtle, for example, that simply refuses to eat. These uncooperative individuals should be released, too.

As to what animals to keep, your choice is nearly unlimited. Pond snails are among the least demanding of all, yet they are interesting creatures to observe. The common species require only a quart jar of pond water for a home and a supply of lettuce for food. Given these, they will soon deposit gelatinous egg masses on the glass. The development of the embryos within the eggs is fascinating to follow with a hand lens.

Crayfish are hardy in captivity, and thrive on a diet of aquatic vegetation and chopped meat, fish, and earthworms. Take care not to overfeed, because the unconsumed food will quickly pollute the water. Provide the tank with a few rocks under which the animals can hide. As the crayfish grow, you will get a chance to see them molt their outgrown external skeletons.

Pond leeches are easy to keep, requiring only a weekly feeding on a piece of raw liver, which should be removed from the water as soon as the leeches have had their fill. Leeches are sensitive to light, vibrations, and the presence of food juices and other chemicals in the water. You can devise simple experiments to test their reactions to these stimuli.

The nymphs of dragonflies and damselflies are interesting to observe. Fierce carnivores, they should be liberally supplied with worms, insects, and other live prey. Furnish their quarters with a stem extending up out of the water, and you may be

216

lucky enough to witness a nymph cast its final molt and emerge as an adult.

Diving beetles, giant water bugs, backswimmers, and water boatmen are worth keeping to observe their different modes of swimming and breathing underwater. Keep their containers covered or they will fly away!

Any of the small species of pond fish are worth a try in cool, well-planted aquariums. Sticklebacks are too scrappy to be housed with other species, but their breeding behavior is so interesting that they deserve a tank to themselves. Native fishes do well on the better foods sold for tropical fishes, such as frozen adult brine shrimp, frozen daphnia, and freeze-dried *Tubifex*. The dry powdered fish foods are not satisfactory.

Because of their remarkable life histories, tadpoles are well worth your attention. If you can, collect the egg masses in the early spring and follow their development all the way through to the adult. Tadpoles are heavy eaters and need a diet of algae, canned spinach, or boiled lettuce, occasionally supplemented with a little lean raw meat. The adult frogs or toads require rather specialized care, and are best released.

Among the salamanders, the common newts make fine aquarium inhabitants. Less retiring than most of their relatives, they eagerly snap up tiny bits of meat or earthworm dangled in front of their noses on the end of a thin wire or broom straw.

Of all the pets available commercially, probably none receives as much ill treatment as the millions of baby turtles sold every

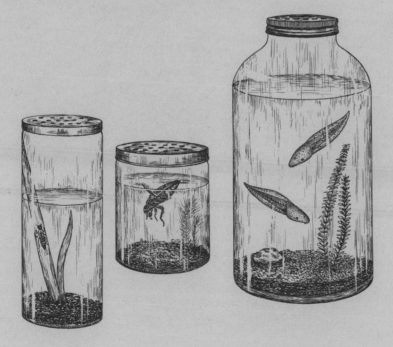

year. Given adequate care, turtles (whether you buy them in a store or catch them yourself) make hardy pets and can grow to considerable size in captivity. The basic requirements are a roomy container with enough clean water for the turtles to submerge completely, and a rock or other object that will allow them to leave the water and bask in the sun or under an electric lamp. Turtles can be fed chopped raw fish, crushed snails, and chopped earthworms. Once a week add a few drops of cod-liver oil to their food. Make sure you buy the plain, unflavored kind. A fifty-cent bottle will last you for years. Some turtles are partly or wholly herbivorous, or become so as they grow older. They should be provided with a more or less constant supply of limp lettuce leaves or other vegetable matter.

The suggestions above are necessarily brief and fragmentary; again, the important thing is to learn all you can about your captives and give them the best care possible. A small collection of well-kept animals is superior to a large collection of sick and dying ones.

Exploring the Microscopic Pond World

The possession of a microscope will quite literally open up an entire new world for you in your explorations of the pond. Each time you place a drop of pond water on a slide and bring it into focus under the microscope, you are venturing into a new, strange, and exciting world. It is an experience that no one ever tires of, no matter how many times he repeats it.

Obviously, the selection of so complicated and specialized an instrument as a microscope poses problems for the novice. If you are contemplating such a purchase, try to seek the assistance of an experienced person—a local science teacher, perhaps, or an established microscope hobbyist. Some of the things to look for—and to look out for—are discussed in the following paragraphs.

Unless your funds are fairly unlimited, you will have to make a choice between an inexpensive new instrument and an older used one. In the first group—microscopes that sell for around fifty dollars or less—the range of quality is from adequate to utterly worthless.

It is possible, for example, to buy for less than twenty dollars a hobby microscope capable of magnifying an object twelve hundred times. A professional instrument that can give good performance in that range of magnification costs between five hundred and fifteen hundred dollars. Something is clearly wrong somewhere. In fact, the combination of a low price tag and high magnification is a sure indication that the instrument is simply a waste of money.

On the other hand, the combination of low price and *low* magnification is usually an encouraging sign. A fifty-dollar microscope probably should not magnify more than two hundred times. This magnification is quite adequate for most pond work; indeed, a great deal of your work will be done at magnifications of from twenty-five to one hundred times.

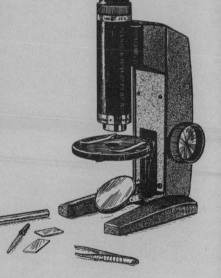

If you live in or near a large city, you might consider purchasing a used microscope; check the classified telephone directory for sources. Properly cared for, a good microscope never wears out, and very fine instruments thirty or more years old are available and are capable of giving excellent service. Prices start at around seventy-five dollars and go up, depending on the age and features of the instrument. Again, bear in mind that for pond work you will not need very high magnification. Try to buy your microscope from a reputable company that offers a guarantee on the instruments it sells.

Whether you settle upon a new microscope or a used one, by all means give it a thorough tryout before you make the purchase. It should yield a flat, crisp image, free of color fringing and distortion. Even at top magnification the image should be bright. All lens surfaces must be in perfect condition and free of chips or scratches. The focusing mechanism should work smoothly and produce a razor-sharp image. All moving parts should have a tight feel to them. A diaphragm to regulate the amount of light entering the microscope is a very desirable feature. A substage condenser is nice to have but not necessary. If you are buying a used microscope, try to avoid one equipped with an oil-immersion objective. This adds a good deal to the cost but little to the usefulness of the instrument for amateur work.

Once you have acquired your microscope, read at least one good book directed to the beginning microscopist. Remember that learning to use a microscope is something like learning to drive a car: it is wise to proceed cautiously and take one thing at a time. A good microscope, even a simple, inexpensive one, is a lifetime investment in one of the most absorbing of all pastimes.

An Invitation to Adventure

The foregoing pages have described a few of the multitude of activities open to the person with enough curiosity to seek out a pond and try to puzzle out some of the many riddles it poses. If you are that kind of person, many questions will probably occur to you as you begin to explore your pond, and you will go to books on aquatic biology for the answers. You may be surprised to find out that many of your questions do not yet have answers.

Even so small and commonplace a habitat as a pond contains vast realms not yet explored by science. If you are so inclined, there is no reason why you should not undertake to explore some part of this unknown territory yourself. You may make no momentous discovery, and your findings may never appear in a professional journal, but if you possess patience, the ability to work in an orderly fashion, and an active, consuming curiosity, you stand an excellent chance of making a contribution, however small, to the sum total of our knowledge of the life of ponds. And apart from the possibility of furthering science, you are certain to find an inexhaustible source of personal satisfaction and enrichment.

Glossary

Adaptation: An inherited characteristic that improves an organism's chances for survival in a particular *habitat*.

Algae (singular *alga*): The simplest of all green plant forms, having neither roots, stems, nor leaves.

Amphibian: Any of a group of animals that includes frogs, toads, and salamanders. Amphibians have soft, moist skins and are characterized by life cycles in which the *larvae* usually live in water and breathe through *gills*, whereas the adults live on land and breathe with *lungs* but return to the water for *spawning*.

Amphipods: A group of small *crustaceans* with bodies compressed from side to side and legs that can be used both for swimming and walking.

Antenna (plural *antennae*): A feeler; an appendage, usually (but not always) sensory in function, that occurs in pairs on the heads of *crustaceans*, insects, and certain other animals.

Aquatic: Living in fresh water, as opposed to marine. *See also* Terrestrial.

Arthropods: Animals with jointed legs and hard external skeletons. The group includes insects, *crustaceans*, spiders, mites, and other similar organisms.

Bacteria (singular *bacterium*): Simple, colorless one-celled plants, most of which are unable to manufacture their own food. Certain bacteria are of importance in the pond as *decomposers*.

Barbels: Fleshy threadlike sensory structures hanging like whiskers near the mouths of certain fishes, such as the catfish.

Benthos: Plants and animals that inhabit the bottom of the pond.

Bivalve: An animal possessing a two-part hinged shell. The fresh-water clam is a bivalve. *See also* Univalve.

Carnivore: A *predatory* animal that lives by eating the flesh of other animals. *See also* Herbivore; Omnivore.

Carrion: Dead animal remains, a source of food for *scavengers*.

Chlorophyll: A group of pigments that produces the green color of plants; essential to *photosynthesis*.

Chloroplast: A *chlorophyll*-bearing body within a plant cell in which *photosynthesis* takes place.

Cilia (singular *cilium*): Minute hairlike structures serving many purposes in a variety of animal groups. Among the *ciliates*, cilia are used for locomotion.

Ciliate: A single-celled organism that swims by means of coordinated movements of *cilia*.

Cladocerans: A group of small *crustaceans*, the water fleas, which have a pair of large Y-shaped swimming appendages.

Coelenterates: A group of simple animals, mostly marine, but represented in ponds by the hydras and the fresh-water jellyfish.

Commensalism: A relationship between two dissimilar organisms, perhaps benefiting both but not essential to the welfare of either. *See also* Mutualism.

Community: All the plants and animals in a *habitat* that are bound together by *food chains* and other relationships.

222

Consumer: Any living thing that is unable to manufacture food from nonliving substances, but depends instead on the energy stored in other living things. *See also* Decomposers; Predator; Primary producers.

Copepods: A group of minute *crustaceans* that have rounded bodies and a pair of elongated oarlike swimming appendages.

Crustacean: A member of a large group of animals that includes the *copepods, cladocerans, amphipods, ostracods*, and similar organisms. Crustaceans are characterized by more than four pairs of jointed legs, segmented bodies, and hard external skeletons.

Decomposers: Organisms, chiefly *bacteria* and *fungi*, that live by extracting nutrients and energy from the decaying tissues of dead plants and animals. In the process, they release chemical compounds stored in the dead bodies and make them available for use by *photosynthetic* plants.

Detritus: Minute particles of the decaying remains of dead plants and animals; an important source of food for many pond animals.

Diatom: A single-celled *alga* encased in an intricately etched pair of silica shells that fit together like a box and its lid.

Dinoflagellates: A group of one-celled organisms that possess characteristics of both plants and animals. Like plants, dinoflagellates can manufacture food through *photosynthesis*; like animals, some are capable of swimming and of catching prey.

Diurnal: Active during the daylight hours. *See also* Nocturnal.

Drought: A prolonged period when little or no precipitation falls on an area.

Ecological niche: An organism's role in a natural *community*, such as *scavenger* or *primary producer*. The term refers to function, not to the place where the organism is found.

Ecology: The scientific study of the relationships of living things to one another and to their *environment*. The scientist who studies these relationships is an ecologist.

Ecotone: An area of transition between one type of *habitat* and another.

Egg: A female reproductive cell. *See also* Fertilization.

Embryo: A developing individual before its birth or hatching.

Emergents: Plants, such as cattails and bulrushes, that root in the mud underwater and protrude above the surface. *See also* Floaters; Submergents.

Energy cycle: The process through which energy from the sun is passed from one living organism to another. Green plants, the *primary producers*, capture solar energy through *photosynthesis*. It is passed on to *herbivores*, then to one or more levels of *carnivores*, and finally to the *decomposers,* with a great deal of energy lost at each step.

Environment: All the external conditions surrounding a living thing.

Estivation: A prolonged dormant or sleeplike state that enables an animal to escape the rigors of hot, dry weather. As in *hibernation*, body processes such as breathing and heartbeat slow down drastically, and the animal neither eats nor drinks.

Estuary: A tidal river; the portion of a river that is affected by the rise and fall of the tide and that contains a graded mixture of fresh and salt water.

Evolution: The process of natural consecutive modification in the inherited makeup of living things; the process by which modern plants and animals have arisen from forms that lived in the past.

Fertilization: The union of a male reproductive cell (*sperm*) with a female reproductive cell (*egg*).

223

Filter feeder: An animal equipped with special body adaptations for straining *plankton* or minute particles of *detritus* from the water.

Flagellate: A single-celled organism that swims by means of one or more *flagella*.

Flagellum (plural *flagella*): A whiplike structure used for locomotion by a group of organisms called *flagellates*.

Floaters: Plants whose leaves float on the surface of the water. Some floaters, such as water lilies, are connected by leafstalks or stems to roots in the pond bottom. Others, such as duckweed, are not so attached, and the entire plant floats. *See also* Emergents; Neuston; Submergents.

Food chain: A series of plants and animals linked by their food relationships. *Plankton*, a plankton-eating *crustacean*, and a crustacean-eating fish would form a simple food chain. Any one species is normally represented in many food chains. *See also* Food web.

Food web: An interrelating system of *food chains*. Since few animals rely on a single source of food and because no food source is consumed by only one species of animal, the separate food chains in a natural *community* such as a pond interlock to form a food web.

Fungi (singular *fungus*): A group of plants lacking *chlorophyll*, roots, stems, and leaves. Some fungi are of importance in the pond as *decomposers*.

Gill: An organ for breathing underwater. Oxygen dissolved in the water passes through the gill membrane into the blood, while carbon dioxide passes from the blood into the water. Among the *aquatic* insects, gills serve as a medium of exchange between the water and the *tracheal system*, rather than the blood. *See also* Lung.

Habitat: The immediate surroundings (living place) of a plant or animal.

Hemoglobin: A complex pigment that imparts the red color to blood and functions as a carrier of oxygen in the blood stream.

Herbivore: An animal (also known as a first-order consumer) that eats plants, thus making the energy stored in plants available to *carnivores*. *See also* Omnivore.

Hibernation: A prolonged dormant or sleeplike state that enables an animal to survive during the winter months in a cold climate. The heartbeat, breathing, and other body processes of the hibernating animal slow down drastically, and it neither eats nor drinks. *See also* Estivation.

Hormone: A chemical substance produced by a living thing and used to regulate the functioning of body processes.

Host: A living organism whose body supplies food or living space for another organism. *See also* Parasite.

Incomplete metamorphosis: The type of life history, characteristic of certain insects such as the dragonflies and true bugs, in which there is no *pupal* stage. Instead, the immature insect, or *nymph*, undergoes a series of gradual changes to transform into the adult. *See also* Larva; Metamorphosis.

Invertebrate: An animal without a backbone, such as a worm, insect, *crustacean*, mollusk, and other forms, which comprise the bulk of the animal kingdom. *See also* Vertebrate.

Larva (plural *larvae*): An active immature stage in an animal's life history, during which its form differs from that of the adult, such as the "wriggler" stage in the development of a mosquito or the tadpole stage in the development of a frog. *See also* Metamorphosis; Pupa.

Limnetic zone: The open-water region of the pond, beyond the *littoral zone*.

Littoral zone: The region of the pond extending from the shore outward to the point at which most plants no longer grow.

224

Lung: A breathing organ consisting of air sacs lined with moist membranes permeated by minute blood vessels. Oxygen from the air passes through the membranes into the blood, while carbon dioxide passes from the blood into the air. *See also* Gill.

Metabolism: The sum of the chemical activities taking place in the cells of a living organism.

Metamorphosis: A change in the form of a living thing as it matures, especially the transformation from a *larva* to an adult. *See also* Incomplete metamorphosis; Nymph; Pupa.

Molt: To shed a body covering, such as the external skeleton of an insect or *crustacean*.

Motile: Capable of free movement. *See also* Sessile.

Mutualism: A relationship between two dissimilar organisms, benefiting both to the extent that neither lives successfully without the other. *See also* Commensalism.

Nekton: *Limnetic* organisms capable of active swimming, such as the fishes and the turtles. *See also* Plankton.

Neuston: Plants and animals that inhabit the surface of the pond, either beneath or on top of the film itself.

Nocturnal: Active at night. *See also* Diurnal.

Nymph: The immature, preadult form of certain insects, such as the dragonflies, whose life histories are characterized by *incomplete metamorphosis*. *See also* Larva; Metamorphosis.

Omnivore: An animal that eats both plants and other animals. *See also* Carnivore; Herbivore.

Organic: Pertaining to anything that is or ever was alive or produced by a plant or animal.

Ostracods: A group of small *crustaceans* characterized by clamlike *bivalve* shells.

Parasite: A plant or animal that lives on or in another organism, its *host,* and obtains shelter and food from the host's body.

Phoresis: An accidental association between two dissimilar organisms, with possible benefit to one. *See also* Commensalism; Mutualism; Symbiosis.

Photosynthesis: The process by which green plants convert carbon dioxide and water into simple sugar and free oxygen. *Chlorophyll* and sunlight are essential to the series of complex chemical reactions involved.

Phytoplankton: Plant *plankton. See also* Zooplankton.

Plankton: The microscopic and near-microscopic plants (*phytoplankton*) and animals (*zooplankton*) that passively drift or float in the *limnetic zone. See also* Nekton.

Plankton bloom: An explosive increase in the plankton population, resulting from a sudden improvement in growing conditions, such as an increase in available sunlight and nutrients.

Predator: An animal that lives by capturing other animals for food.

Primary producers: *Photosynthetic* plants, which manufacture the food on which all other living things ultimately depend. *See also* Consumer.

Production pyramid: The diminishing amount of organic material present at each successive level along a *food chain.* The decline results mainly from the constant loss of energy through *metabolism* along the food chain. *See also* Pyramid of numbers.

Protista: A term embracing all one-celled organisms, whether plant or animal—the *protozoans,* the *algae,* and the *bacteria.*

Protozoan: A simple, one-celled animal such as *Amoeba* or *Stentor.*

225

Pupa (plural *pupae*): The relatively inactive stage in certain insects, such as the mosquitoes and midges, during which a *larva* undergoes *metamorphosis* into an adult form. *See also* Incomplete metamorphosis.

Pyramid of numbers: The normally declining number of individuals at each successive level on a *food chain. See also* Production pyramid.

Scavenger: An animal that eats the dead remains and wastes of other animals and plants. *See also* Predator.

Sessile: Permanently attached to a surface; sedentary. *See also* Motile.

Sexual reproduction: Formation of a new generation through the union of female germ cells (*eggs*) and male germ cells (*sperms*). *See also* Fertilization.

Spawn: To shed reproductive cells. The term refers to animals, such as fishes, that shed *eggs* and *sperm* directly into the water.

Sperm: A male reproductive cell. *See also* Fertilization.

Spiracle: An opening for breathing, such as the external opening to an insect's *tracheal system* or the opening through which a tadpole expels water as it breathes.

Stomate: A microscopic opening in the surface of a leaf that allows gases to pass in and out.

Submergents: Plants, such as wild celery

and water milfoil, which grow wholly underwater. *See also* Emergents; Floaters.

Succession: The gradual replacement of one *community* by another.

Surface tension: A property of liquids that causes the surface of a liquid to act as an elastic film. Surface tension results because molecules of liquid have a stronger attraction for each other than they do for the air molecules above them.

Symbiosis: An association of two dissimilar organisms in a relationship that may benefit one, both, or neither. *See also* Commensalism; Mutualism; Parasite; Phoresis.

Terrestrial: Living on land. *See also* Aquatic.

Territory: An area defended by an animal against others of the same species. It is used for breeding, feeding, or both.

Tracheal system: In insects and certain of their relatives, a system of minute branching air tubes called tracheae (singular *trachea*) that distributes air throughout the body, bringing in oxygen and carrying away carbon dioxide. *See also* Spiracle.

Univalve: Possessing a single, unhinged shell. Snails and limpets are univalve. *See also* Bivalve.

Vertebrate: An animal with a backbone protecting a nerve cord. The vertebrates comprise fishes, *amphibians*, reptiles, birds, and mammals. *See also* Invertebrate.

Zooplankton: Animal *plankton. See also* Phytoplankton.

226

Bibliography

FRESH-WATER BIOLOGY

BENNETT, G. W. *Management of Artificial Lakes and Ponds*. Reinhold, 1962.

BROWN, E. S. *Life in Fresh Water*. Oxford University Press, 1955.

COKER, ROBERT E. *Streams, Lakes, Ponds*. University of North Carolina Press, 1954.

EDMONDSON, W. T. (Editor). *Freshwater Biology*. Wiley, 1965.

KLOTS, ELSIE B. *The New Field Book of Freshwater Life*. Putnam, 1966.

MACAN, THOMAS T. *Freshwater Ecology*. Wiley, 1963.

MACAN, THOMAS T., and E. B. WORTHINGTON. *Life in Lakes and Rivers*. Collins, 1951.

MORGAN, ANN HAVEN. *Field Book of Ponds and Streams*. Putnam, 1930.

NEEDHAM, JAMES G., and J. T. LLOYD. *Life of Inland Waters*. Comstock, 1916.

NEEDHAM, JAMES G., and PAUL R. NEEDHAM. *A Guide to the Study of Fresh-water Biology*. Holden-Day, 1962.

POPHAM, EDWARD J. *Some Aspects of Life in Fresh Water*. Harvard University Press, 1961.

RUTTNER, FRANZ. *Fundamentals of Limnology*. University of Toronto Press, 1963.

WELCH, PAUL S. *Limnology*. McGraw-Hill, 1952.

ANIMAL LIFE

BARNES, ROBERT D. *Invertebrate Zoology*. Saunders, 1963.

CARR, A. *Handbook of Turtles*. Comstock, 1963.

CARTHY, J. D. *Animal Navigation*. Scribner, 1956.

CHU, HUNG-FU. *How to Know the Immature Insects*. William C. Brown, 1949.

CONANT, ROGER. *A Field Guide to Reptiles and Amphibians of Eastern North America*. Houghton Mifflin, 1958.

CORBET, PHILIP S. *A Biology of Dragonflies*. Quadrangle Books, 1963.

CURTIS, BRIAN. *The Life Story of the Fish*. Dover, 1949.

DOWDESWELL, W. H. *Animal Ecology*. Harper & Row, 1961.

EDDY, S. *How to Know the Freshwater Fishes*. William C. Brown, 1957.

FRAENKEL, GOTTFRIED S., and DONALD L. GUNN. *The Orientation of Animals*. Dover, 1961.

GRAY, JAMES. *How Animals Move*. Penguin, 1959.

HERALD, E. S. *Living Fishes of the World*. Doubleday, 1965.

HYLANDER, CLARENCE J. *Fishes and Their Ways*. Macmillan, 1964.

JAHN, T. L. *How to Know the Protozoa*. William C. Brown, 1949.

JAQUES, H. E. *How to Know the Insects*. William C. Brown, 1947.

LUTZ, FRANK E. *Field Book of Insects*. Putnam, 1935.

NEEDHAM, JAMES G. *Culture Methods for Invertebrate Animals*. Dover, 1937.

PENNAK, ROBERT W. *Fresh-water Invertebrates of the United States*. Ronald Press, 1953.

PETERSON, ROGER TORY. *A Field Guide to the Birds*. Houghton Mifflin, 1947.

POUGH, RICHARD H. *Audubon Water Bird Guide*. Doubleday, 1951.

ROMER, ALFRED S. *The Vertebrate Story*. University of Chicago Press, 1962.

STEBBINS, ROBERT C. *Amphibians and Reptiles of Western North America*. McGraw-Hill, 1954.

USINGER, ROBERT L. (Editor). *Aquatic Insects of California*. University of California Press, 1956.

WIGGLESWORTH, V. B. *Principles of Insect Physiology*. Dutton, 1965.

WRIGHT, ALBERT H., and ANNA A. WRIGHT. *Handbook of Frogs and Toads of the United States and Canada*. Comstock, 1933.

PLANT LIFE

BOLD, H. C. *The Plant Kingdom*. Prentice-Hall, 1965.

FAIRBROTHERS, DAVID E., and others. *Aquatic Vegetation of New Jersey*. Rutgers Extension Bulletin No. 382.

FASSETT, NORMAN C. *Manual of Aquatic Plants*. University of Wisconsin Press, 1957.

HOTCHKISS, NEIL. *Pondweeds and Pondweedlike Plants of Eastern North America*. U.S. Department of the Interior, 1964.

MUENSCHER, W. C. *Aquatic Plants of the United States*. Comstock, 1944.

PRESCOTT, G. W. *How to Know the Fresh Water Algae*. William C. Brown, 1964.

ECOLOGY

BENTON, ALLEN H., and WILLIAM E. WERNER, JR. *Field Biology and Ecology*. McGraw-Hill, 1966.

BUCHSBAUM, RALPH, and MILDRED BUCHSBAUM. *Basic Ecology*. Boxwood Press, 1957.

ODUM, EUGENE P. *Ecology*. Holt, Rinehart and Winston, 1963.

ODUM, EUGENE P., and HOWARD T. ODUM. *Fundamentals of Ecology*. Saunders, 1959.

GENERAL READING

BROWN, VINSON. *The Amateur Naturalist's Handbook*. Little, Brown, 1948.

CARRIGHAR, SALLY. *One Day at Teton Marsh*. Knopf, 1947.

DE BEER, G. *Atlas of Evolution*. Nelson, 1964.

DOBELL, CLIFFORD. *Antony van Leeuwenhoek and His "Little Animals."* Russell & Russell, 1958.

PALMER, E. LAURENCE. *Fieldbook of Natural History*. McGraw-Hill, 1949.

SIMPSON, GEORGE GAYLORD, and WILLIAM S. BECK. *Life: An Introduction to Biology*. Harcourt, Brace & World, 1965.

Illustration Credits and Acknowledgments

AUTHOR'S ACKNOWLEDGMENT: *A book of this sort is a compendium from many sources, including books, journals, and my colleagues at the University of Delaware and elsewhere. My wife and children have been with me on countless trips to ponds throughout the country. But it is perhaps to my students that I owe the greatest debt of gratitude, for they have worked with me for twenty years on local ponds and have opened my eyes to much that I could not have otherwise seen.*

The publisher also wishes to thank F. C. Gillette, former Chief, Division of Wildlife Refuges; C. Gordon Fredine, Chief, Division of International Affairs of the National Park Service; and the park superintendents and refuge managers who responded in detail to a questionnaire on ponds in areas administered by the Department of the Interior. The publisher also expresses appreciation to William Perry, O. L. Wallis, and M. Woodbridge Williams of the National Park Service for their assistance in reading the manuscript or locating photographs. Finally, the publisher gratefully acknowledges permission to reprint the passage on page 161 from Clifford Dobell (Ed. and Transl.), Antony van Leeuwenhoek and His "Little Animals," Russell & Russell, New York, 1958.

228

Index

[Page numbers in **boldface** type indicate reference to illustrations and maps.]

Adaptations, 102, 116, 146, 174, 178, 183, 222
  color, 149
  of floating leaves, 44, 46
  respiratory, 124–127, 131–133
  seasonal, 97
  (*See also* Evolution)
Air (*see* Atmosphere)
Air sacs, 119, 132
Airspaces, 102, 123, 168
Alderflies, **66**
Algae, 20, 59, 68, 82, 89, 102, 111, 134, **151**, 171, 175, **184–185**, 192, 222
  habitat of, 162, 168
  mutualism of, 153, 182
  as primary producers, 24
  reproduction of, 32, 53, 106, 184
  swimming, 185, 191
  (*See also* Desmids; Diatoms; Flagellates)
Alligators, **12**, 204
Amoeba (protozoans), 101, **122**–123
Amphibians, 83, 86, 100, 188, 222 (*see also* Frogs; Newts; Salamanders; Toads)
Amphipods, 41, 59, 62, 128, **151**, 222
  nighttime activity of, 77
  as.scavengers, 157
  winter activity of, 95
Anabaena (blue-green algae), 184
Anacharis (waterweed), 103–104
Animals, 16, 20, 23–24, 57, 95, 108–110
  evolution of, 99–102
  one-celled (*see* Protozoans)
  zonation of, 50–51
  (*See also* Breeding; Plankton; Zooplankton; *specific animals*)
Antennae, 121, 171, 182, 194, 222
Aquariums, 103, 117, 181, 214–217
Aransas National Wildlife Refuge (Tex.), 202
Asplanchna (rotifers), **192**
Asterionella (diatoms), **107**, **193**
Atmosphere, 95, 105, 125
  oxygen in, 82, 123–124
  temperature of, 57, 65, 89
Autumn, 80–81, 86, 88–89, 95, 103
Azolla (water velvet), 106

Backswimmers, 65–66, **115**, 118, 168, **171**
  breathing of, 124–125, 127
Bacteria, 33, 66, 80, 157, 162, 182, 222
  bottom-dwelling, 179, 181
  in food chains, 23–25, 150
  removal of, 20
  reproduction of, 53
  in shore zone, 196, 199
Barbels, 179, 222
Bass, 24, **42–43**, 62, 67, 138, 142, 195
Beaches, 16, 196–197, 199
Beavers, 16–19, 52, 75, 86, 95, 105, 108–109, 134
Beetles, 119, 166, **169**, 196–197 (*see also* Water beetles; *specific beetles*)
Benthic regions (*see* Bottoms)
Benthos, 20, 222

Birds, 41, 65, 77, 81, 86, 88, 144
  as consumers, 24, 26, 134
  evolution of, 100
  flightless, 109
  predatory, 24, 26, 28, 72, 77, 82, 108, 138–**139**
  as restorers of pond food, 83, 88
  as seed carriers, 32
  (*See also* Waterfowl; *specific birds*)
Bitter Lake National Wildlife Refuge (N.M.), 202–203
Bivalves, 194, 222 (*see also* Clams; Mussels)
Bladders, excretory, 153
  swim, 115
  as traps, 164–**165**
Bladderwort, 32, 46, 89, 105, **164–165**
Blood, 127, 131–132, 146, 156, 193
Blood vessels, 89, 130–131
Bloodworms, **126**
Blue-green algae, 106, 162, 168, **184**
Body temperature, regulation of, 89
Bottoms, 20, 88–89, 103, 120–121, 144, 183, 188, 194
  decomposers on, 24
  detritus on, 25, 157, 178–179, 192–193
  drying of, 79–80
  false, 35
  as habitats, 177–179, 181
  immature insects on, 146, 159
  springtime activity on, 59, 62, 66
  in succession, 28–29, 32–34, 50
  winter state of, 95, 97
  (*See also* Sediment)
Breathing, 123–125, 127–133, 155, 159
  (*see also* Gill breathing; Oxygen; Suffocation)
Breeding, 20, 38, 52–53, 61, 118–119, 159, 168, 192, 194, 196
  autumnal, 88
  of crayfish, 129
  energy for, 27
  of fish, 32, 41, 66–67, 134, 195
  of fisher spiders, 149
  of parasites, 153, 155
  role of surface film in, 176–177
  springtime, 59, 62, 65–66, 110, 134, 195
  summertime, 66–67, 72, 74–75, 80–81, 88
  of turtles, 113
  of water striders, 173
  winter, 97
Bristle worms (*Chaetogaster*), **169**, 179
Bristles, **116**, **119**, **128**, 192, 194
Bryozoans, 88, 168–**169**, 182–**183**
Bubbles, 32, 59, 79, 95, 116, 121, 124–125, 127
Budding, asexual, 88, 103–104
Bullfrogs, 66, 110, 141, **143–144**, **151**
Bullheads, 41, **43**, 46, 53
Bulrushes, 34, **50–51**
Buoyancy, 103, 114–116, 119, 132
Burrowers, 51, 53, 109, **129**
  bottom habitat of, 34, 178–179
  insect, 108, 146, **196**–197
  worms as, 79, 95, 196–197

Caddisflies, 119, 127, 152, **169**
Canada geese, **74–75**
Cape Hatteras National Seashore (N.C.), 203
Carbon dioxide, 23, 48, 82, 102, 125, 130, 133, 153
Carolina Sandhills National Wildlife Range (S.C.), 203
Caterpillars, lily-leaf, **167**, **169**
Catfish, 43, 179
Cattails, **22**, 32, 34–**35**, **50–51**, 53, **108**
Cellular respiration, 123
Cellulose, 185, 191
Chaetogaster (bristle worms), **169**, 179
Chaetophora (green algae), 107
Chaoborus (phantom midge larvae), 25, 119, 149
Chlorophyll, 23, 222
Chloroplasts, 23, 222
Chubsuckers, 46
Cilia, 122–123, 128, 162, 181–182, 192
Ciliates, 122–123, 152–153, 222
Cladoceran water fleas, 25, 95, 97, 121, **149**, **151**, 175, 182, **193–194**, 222
Clams, 80, 128, 155–156, 168–**169**, **180**–181 (*see also* Mussels)
Clarence Rhode National Wildlife Range (Alaska), 203
Climate (*see* Atmosphere; Drying; Rainfall; Seasons; Wind)
Closterium (desmids), **184–185**
Coelenterates (*see Hydra*)
Columbia National Wildlife Refuge (Wash.), 203–204
Commensalism, 152, 159, 222
Conochilus (rotifers), 192
Consumers, 24–28, 46, 80, 82, 107–108, 133, 153, 223
Contractile vacuoles, 101
Copepods, 52, 119, 133, **150**, 152–153, 162, 179, 182–183, 197, 223
  breathing of, 129
  estivation of, 80
  in food chains, 26–27
  as intermediate hosts, 155
  nighttime activity of, 77, 149
  population of, 193, 199
  reproduction of, 59
  as swimmers, 121
Corixids (water boatmen), 51, 65–66, 79, **118**–119, 125, 127, 153
Crane flies, **66**
Craspedacusta (jellyfish), **120–121**
Crawlers, 108, 121
Crayfish, 34, 42, 50, 76, 84, 108–109, 128–**129**, 141, 144–145
  burrowing, 53, 129, 179
  as scavengers, 150–151, 157
Cristatella (bryozoans), 168
Crustaceans, 32, 52, 72, 77, 95, 117, 128, 179, 193–194, 223
  bottom-dwelling, 41, 194
  evolution of, 100–101
  in food chains, 20, 26, 146–147
  as intermediate hosts, 155
  phoresis of, 152
  as scavengers, 157

Crustaceans, as swimmers, 120–121
(See also Amphipods; Copepods;
Crayfish; Isopods; Ostracods;
Shrimp; Water fleas)
Currents, 66, 106, 114–115, 128–129,
171, 177
Cyclops (copepod shrimp), 52, 183, **193**
Cysts, 32, 53, 80, 86, 155–156, **191**
Cytoplasm, 101, 107, 123

Damselflies, 119, **153**, 156, 166
nymphs of, 59, 79, 127
reproduction of, 168, **176**
Daphnia (water fleas), **95**, 97, 121,
193–194
Darters, 115
Decomposition, 24–25, 66, 97, 106, 150,
157, 179, 181, 223
in succession, 28–29, 32–33, 52–53
Deer, **136–137**
Desmids, **23**, 59, 133, 162, 168, **184–185**
Detritus, 25, 118, 123, 128, 150, 152,
157, 159, 162, 168, 178–179, 181,
192–193, 223
Diatoms, 28, 59, 95, 107, 133, 175, **193**,
223
habitats of, **162**, 168
reproduction of, 184
Dineutes (whirligig beetles), 41, **116**,
174–175
Dinobryon (flagellates), 191
Dinoflagellates, 191, 223
Diving beetles, 115–118, **150–151**, **171**
breathing of, 124–125, 127
as predators, 117, 146
Dodder, **81**
Dolomedes (fisher spiders), 41, **148**–149
Dormant states, 59, 79–80, 86, 89, 97
Dragonflies, 80, **115**, 119, **144**, **166**, 168,
176–**177**
breathing of, 127–128, 155
as burrowers, 179
as predators, 146–147, 166
springtime activity of, 59, 62–65
Dredges, **188**
Drying, 153, 155, 223
seasonal (droughts), 52, 57, 79–80
in succession, 28–29, 51–53, 55
Ducks, **38–39**, 41, 46, 50, 60–62, 84,
105, **132–133**, 162
in food chains, 24–25, 134, 138, 145
summertime activity of, 72
winter activity of, 97
Duckweed, **44**, 59, 105–106, 134,
161–162, 171
Dugesia (planarian flatworms), 129
Dune ponds, **15–16**
Dytiscus (diving beetles), **115–117**, 124

Ecological niches, 159, 183, 223
Ecotones, 196, 223
Egrets, 72
Emergents, 35, 41, 44, 51–52, 59,
101–103, 145, 150, 176, 223
in autumn, 88–89
in littoral zone, 36, 134
summertime activity among, 79–81
(See also specific emergents)
Energy, 23, 25, 77, 99, 133, 178, 223
loss of, 82–83
transfer of, 26–28, 150–151
Estivation, 79–80, 223
Euglena (flagellates), 107, **122**, 185
Eutrophic ponds, **48**
Evaporation (see Drying)
Everglades National Park (Fla.), **12**, 204
Evolution, 99–102, 111, 113, 223 (see also
Adaptations)
Eyes, 19, **64**, 77, 145, 149, 193, 206–207

Eyes, eyespots as, 185, 191–192
of frogs, **69–70**
of turtles, 79, **113**, 156
of whirligig beetles, **174–175**

Fanwort, 46, 134
Ferrissia (snails), 168
Filter feeders, 80, 181–183, 224
Fins, 114–115, 155
First-order (primary) consumers, 24, 26,
80, 82, 107–108, 133
Fish, 9, 32, 46, 61, 72, 97, 108–110, 132,
189, 206–207
breathing of, 79, 130–131
breeding of, 32, 41–43, 66–67, 110,
134, 195
evolution of, 100–101
in food chains, 20, 24–26, 28, 134,
138–147, 149
as hosts, 155–156, 158
larval, **195**
limnetic, 20
schools of, 67
summertime activities of, 76–77, 79–81,
84
as swimmers, 114–115
vision of, 206–207
winter activities of, 95
(See also specific fish)
Fish and Wildlife Service, 12
Fisher spiders, 41, **148**–149
Flagella, 122, 185, 191, 224
Flagellates, 59, 122, 133, 153, **185**,
190–192, 224
Flatworms, 41, 79, 157, 159, **169**, 171, 179
breathing of, 129
breeding of, 52, 62
parasitic, 155
sponge-dwelling, 152
Flies (maggots), 59, 77, 123, 157, **171**
Floating heart, 44, 166
Floating plants, 44, 46, 52, **104**, 106,
150, 192, 224
as microhabitats, 166–168
(See also specific floating plants)
Flowers, 46, 57, 59, 81, 88, 102, **104**–105
as microhabitats, 164, 168
(See also Pollen)
Food (see Detritus; Food chains;
Nutrients; Parasites; Predators;
Symbiosis)
Food chains, 20, 23–28, 145–151, 224
decomposers in, 24–25
reptiles in, 138, 140–141, 145
Food webs, **150–151**, 224
Forests, 19, 55
Freezing (see Ice)
Frogs and tadpoles, 33–35, 41, 46, 53,
61–62, 109–110, 117, 134
in food chains, **24**, 26, 138, 140–141,
143–146, 149, 151
hibernation of, 89
as hosts, 156
snorkels of, 123
summertime activities of, 66–72, 76–77,
79–84, 88
Fungi, 24–25, 53, 179, 181, 224

Gas sacs, 119, 132
Gastrotrichs, 192–193, 197
Geese, 50, **74–75**, 134
Gemmules, **88**
Gerris (water striders), 172
Giant water bugs, **146**
Gill bailers, 129
Gill breathing, 67, 79, 89, 95, 100, 110,
127–131, 224
of mollusks, 180–181
physical substitute for, 124–125, 127
of tadpoles, 68, 70–71

Gill filaments, **130**, 155
Glacial ponds, 16
Glochidia, 155–156
Gloeotrichia (blue-green algae), 184
Glowworms, 196
Grand Teton National Park (Wyo.), **13**
Grasses, 34–35, 55, 72, 103, 134
Great Swamp National Wildlife Refuge
(N.J.), 204
Grebes, 46, 50, 62, 72, 109, 132–133
Green algae, 106–107, 168, 175, **184–185**
Gulf Island National Wildlife Refuge
(Miss.), 204
Gyraulus (snails), 168
Gyrinus (whirligig beetles), 116

Habitats, 33, 166–168, 170–179,
181–183, 224
microhabitats, 161–162, 164, 183
(See also specific habitats)
Hairs, 102, 164, 197 (see also Bristles;
Cilia; Water-repellent hairs)
Harpacticoids, 183, 197
Haustoria, 81
Hawks (ospreys), 24, 26, 28, 82, 108,
138–**139**
Heat energy, 27–28 (see also Sunlight)
Heliozoans, 123
Hemoglobin, 127 132, 224
Herons, 20, 34–35, **72–73**, 82, 97, 141
Hibernation, 86, 89, 97, 224
Hornwort, **46**, 105, 134
Hydra (coelenterates), **88**, 121, 129,
152–153, 171
habitat of 161–163, **169–170**
Hydrodictyon (water net), **184–185**

Ice, 16, 30, 57–59, 89, **94–95**, 97, 114,
191
Insects, 9, 41, 50–52, 68, 72, 76–77,
80–83, 108–109, 168
bottom-dwelling, 20, 146, 159
breathing of, 123–125, 127
evolution of, 100
in food chains, 24–26, 134, 144–146,
149
phoresis of, 152
pioneer, 29, 32–33
pollinating, 59, 80, 105
preying on insects, 62, 65, 80, 117–118,
146–147, **156**, 173, 197
as swimmers, 108, 116, 118–119
(See also Larvae; Nymphs; specific
insects)
Invaders, 35–37, 50–52, 103
Isopods, 41, 59, 62, 157, 179
gills of, 128
nighttime activity of, 77
winter activity of, 95

Jellyfish, **120–121**

Keratella (rotifers), 192
Kidneys, evolution of, 100–101
Kingfishers, 20, 72, 76, 82, **151**

Labia, dragonfly, 146–147
Lacreek National Wildlife Refuge (S.D.),
204
Lakes, 10–11, 14–16, 28, 36, 42, 55, 99,
120
Larvae, insect, 42, 79–80, 157, 177, 189,
224
breathing of, 79, 123–128
burrowing, 127, **196**
in food chains, 20, 25–26, 146, 149–151

Larvae, habitats of, 162, **167**–168,
    196-197
  as hosts, 152
  parasitic, 153
  springtime activity of, 59
  as swimmers, **115**, 119
  winter activity of, 95
Leeches, 53, 79–80, **114**–115, 156, **158**,
    189
Leeuwenhoek, Anton van, 161–162, 191
Leptodora (water fleas), 25, **149**, **151**, 194
Lice, 155, **166**, **170**
Light, 99, 162, 184, 196, 206–207
  sensory detection of, 127, 153, 173, 185
  (*See also* Night; Photosynthesis;
    Sunlight)
Limnetic zone, 20, 46, 224
Littoral zone, 20, 28–29, 34–36, 41, 48,
    106, 109, 141, 224
  of aging ponds, 52–53
  bullfrogs in, 144
  food supply in, 134
  harpacticoids in, 183
  invaders from, 36, 52, 103
  oxygen supply in, 123, 196
  pioneers of, 32
  sandy, 196–197, 199
Liverwort, 44, 106
Loons, 132–133
Loosestrife, 35–**36**, 41, **81**, 88, 103
Lungs, 100, 123–124, 132, 141, 225

Maggots, 123, 157, **171**
Mallards, **38**–**39**, 41, 46, 50, 72, 97, **150**
Mammals, 24, 72, 83, 88, 134, 144
  breathing of, 124
  evolution of, 100
  as hosts, 115, 156
  (*See also specific mammals*)
Man-made ponds, **16**, **29**, 32
Marshes, 34–35, 52, 58, 109, **129**
Mayflies, 26, **66**, 81, 119, 127–128, 179
Meadows, 19, 28–29, 36, 55
Meanders, 15, 29, 32
Medusas, 120–121
Mergansers, 46, 60–62, 97, 132–133
Metabolism, 27–28, 103, 111, 133, 153
Metamorphosis, **68**–**71**, 155–156, 224–225
Microcystis (blue-green algae), 184
Microscopes, **189**, 219–220
Microscopic organisms (*see* Bacteria;
    Phytoplankton; Protozoans)
Midges, 62, 76, 82, **126**, 146, 179, 196
  breeding of, 177
  in food chains, 26, 149–151
  habitats of, 168, 171
  larval breathing of, 79, 127–128
  as swimmers, **115**, 119
Migration, 32, 53, 80, 86, 121, 153, 155
Minerals, 14, 23, 48, 58, 83
Minnows, 26, 41, 53, 67, **142**–**143**, **151**,
    170–171
Moles, 95, 108–109
Moose, 134–**135**
Mosquitoes, 76, 80, 123, **170**, 176–177
Moss, 35, **41**, 102, 106
Mucus, 53, 103, 180
Mud Pond, **50**–**51**
Muskellunge, 145
Muskrats, 34, 52, 62, 82, 84, 105, **108**–109
  in food chains, 25, 134, 150
  trapping of, 108
  winter activity of, 95
Mussels, 46, 108, 128, 153, 155–156, 179,
    181
Mutualism, 152–153, 159, 182, 225

National Bison Range (Mont.), **12**
National Park Service, 12, 202

Nekton, 20, 225
Nets, **25**, 149, 162, **187**–**189**, 199, 212
Neuston, 20, 170, 225
Newts, **110**
Night, 32, 77, 84, 102, 119, 149, 171,
    196
Nutrients, 30, 32–33, 48, 81–83, 102
  accumulated in winter, 57
  decayed, 24, 28, 181
  loss and restoration of, 82–83, 88–89
  pollution and, 106
  (*See also* Minerals)
Nymphs, insect, **63**, 76, 119, 179, 196,
    225
  breathing of, 127–128, 155
  in food chains, 26, 146–147
  springtime activity of, 59, 62–63, 66
  summertime activity of, 79–80

Oceans, 15–16, 99, 114, 192–193, 199
Oedogonium (green algae), 107
Oligochaetes (segmented worms), **150**,
    152, 178–179, 196–197
Oligotrophic ponds, **49**
Olympic National Park (Wash.), 205
Opercula, 130–**131**
Oscillatoria (blue-green algae), 106, **184**
Ospreys, 24, 26, 28, 82, 138–**139**
Ostracods, 77, 80, **169**, 182, **194**, 225
Otters, 82, 108–**109**, 124, 138, **151**
Owls, 77, 108
Oxbow lakes, **14**–15
Oxygen, 66, 102, 123–125, 127–129,
    131–133, 153, 180
  absorbed in hibernation, 89
  atmospheric, 82, 123–124
  bubbles of, 32, 59, 79, 95, 116,
    124–125, 127
  plant consumption of, 77, 102
  in plant stems, 103, 168
  supply of, 33, 42, 48, 53, 79, 95, 106,
    131, 149, 162, 177–178, 196
  for swimming, 115–116, 119
  symbiosis and, 152–153

Painted turtles, 41, 89, 134
Paramecium (protozoans), 77, 101,
    152–153
Parasites, 81, 111, 115, 128, 153,
    155–156, 158–159, 225
Pectinatella (bryozoans), 88, 182–**183**
Pediastrum (green algae), **184**–185
Pepperwort, 106
Perch, **42**, 46, 76, 79, 155
Peritrichs, 152, 161–**162**, 182
Phacus (flagellates), **185**
Phantom midges, 25, **119**, 149–151
Phoresis, 152, 159, 225
Photosynthesis, 23, 46, 59, 81, 99, 153,
    225
  in pond's depths, 177–178
  stored products of, 102–103, 123
Physa (snails), 168–**169**
Phytoplankton, **23**, 25–27, 59, 66,
    106–107, 133, **150**, 181, 183–185,
    191–192, 225
  bloom of, 134
  evolution of, 101
  as light shield, 177
  in summer, 80
  (*See also* Algae)
Pickerel, **34**, 41, 46, 67, 142, **145**, **151**
Pickerelweed, **40**–41
Pied-billed grebes, 46, 62, 109, **132**–**133**
Pike, 145
Planarians (*see* Flatworms)
Plankton, 20, 32, 46, 183–185, 188,
    191–195, 225
  in aging ponds, 53

Plankton, animal (*see* Zooplankton)
  bloom of, 57–59, 89, 106–107, 134,
    225
  density of, 46, 48
  in food chains, 24–26
  habitats of, 161–162, 164, 166–168
  plant (*see* Phytoplankton)
  in winter, 95, 97
  in young ponds, 29
Plants, 10, 57, 80, 101–107, 146, 152,
    156, 168
  carnivorous, 32, 164
  littoral, 20, 34
  in oxbow lakes, 15
  pioneer, 28–29, 32–33
  pollination and reproduction of, 32–33,
    46, 55, 59, 81, 88–89, 103, 105,
    184, 191
  as primary producers, 23–25, 28, 80,
    88
  zonation of, 34–35, 41, 44, 46, 50–51,
    103
  (*See also* Detritus; Emergents; Floating
    plants; Invaders; Photosynthesis;
    Plankton; Submergents; *specific plants*)
Pleurosigma (diatoms), **184**
Pollen and pollination, 59, 80, 83, 105,
    171
Ponds, defined, 10–11, 14
Pondweed, 44, **46**, **102**, 105, 134
Potholes, 16, **31**
Predators, 53, 80, 108, 116–118, 144–146,
    152, 156, 159, **196**–**197**, 225
  benthic, 178
  microhabitats of, 162
  on plankton, 25–26, 192
  reptiles as, 138, 140–141
  spiders as, 149
  (*See also specific predators*)
Primary (first-order) consumers, 24, 26,
    80, 82, 107–108, 133
Primary production, 23–26, 28, 80, 88,
    225
Protista, 185, 225
Protozoans, 77, 80, 101, 129, 225
  bottom-dwelling, 179, 181
  habitats of, 161–**162**, 168, 199
  planktonic, 192
  reproduction of, 32, 53
  as swimmers, 120, 122–123
  symbiotic, 152–153
  (*See also specific protozoans*)
Pseudopodia, 122
Pupae, insect, 76, 123, **170**, 226
Pyramid of numbers, 28, 225–226
  (*See also* Food chains)

Quarry ponds, 16, 120

Raccoons, 82–85
Rainfall, 30–31, 57, 80, 82 (*see also*
    Drying)
Red-winged blackbirds, 41, 81
Reed grass, 34–35, 72, 103
Reproduction, 86, 226
  asexual (vegetative), 88, 103–104
  plant pollination, 32–33, 46, 55, 59,
    81, 88–89, 103, 105, 184, 191
  (*See also* Breeding)
Reptiles, 86, 100, 124, 138, 140–141, 156
  (*see also* Snakes; Turtles)
Reservoirs, 16, 120
Respiration (*see* Breathing; Gill breathing;
    Oxygen; Suffocation)
Rivers, 14–16, 28, 42, 99, 114
Rotifers, 77, 95, 151–153, 182, **192**–**193**
  habitats of, 161–**162**, 164, 168–**169**,
    197, 199
  reproduction of, 32, 53, 59, 88, 192

Roundworms, 152, 155, 178, 196–197
Rove beetles, 119, **171**
Rushes, 32, 34–35, 41, **50–51**, 53, 102

Salamanders, 52–53, 89, 108, 140
Salt water, 16, 99–100
Savannah National Wildlife Refuge
  (S.C.), 205
Scavengers, **24**–25, 77, 80, **150–151**, 157,
  173, 226
  turtles as, 140, 159, 179
Seasons, 52–53, 57–97
  dry, 52, 57, 79–80
  and water temperature, 11, 14–15, 25,
    58, 62, 68, 79, 89
  (See also specific seasons)
Second-order consumers, 24–28
Sedges, 34–35, 53
Sediment, 30, 32–33, 46, 52, 66, 115, 181
  accumulation of, 16, 19, 28, 33–34, 52,
    89, 97, 178
  and animal zonation, 51
  in false bottoms, 35
  in shore zones, 196
Sensory organs, 76, 127, 133, 149, 153,
  173, 179 (see also Eyes)
Sessile organisms, 164, 226
Sherburne National Wildlife Refuge
  (Minn.), 205
Shiners, **42**, 46, 53, 72, 81
Shore zone (see Littoral zone)
Shrimp, 52, 80, 183, 193
Shrubs, 34, 36, 52, 55, 144
Silica cases, **107**, 184
Silt (see Sediment)
Silver Spring Lake, **36–37**
Siphons, **180–181**
Sisyra (spongillaflies), **128**, 153
Skin, waterproof, 89, 100
Skunk cabbage, **58–59**, 107
Snails, 32, 41, 53, 62, 67, 117, 123, 147,
  156–157, 168–171, 179, 189
  diet of, 20, 134
  estivation of, 80
  evolution of, 100
  phoresis of, 152
  vertical migration of, 121
Snakes, **34**, 79, 83, 88–89, 109, 115,
  140–145
Snapping turtles, 110, 140, 156, 159, 179
  in food chains, 26–28, 138, 140–141,
    150
Soil, 29, 33–34, 53, 58, 80–82, 196
Spatterdock (yellow water lilies), 44–**45**,
  72, 81, 88, 105, 135–138, 145
Sphagnum moss, 35, **41**, 106
Spiders, 41, 100, **148**–149, 155
Spike rushes, 32, 34–35, 41, 103
Spines, 164, 184–185, 191–192
Spiracles, 124, 226
Spirogyra (green algae), 107
Spirostomum (protozoans), **122**
Sponges, **88**, 128, 152–153, 168–**169**, 182
Spongilla (sponges), **88**, 153, 168–**169**
Spongillaflies, **128**, 153
Spring, 52, 57–59, 81, 97, 191
  autumn contrasted to, 88–89
  breeding in, 59, 62, 65–66, 110, 134,
    195
Springtails, 166–167, **170**
Statoblasts, **88**, 183
Stentor (protozoans), **163**, **182**
Stenus (rove beetles), 119, **171**
Sticklebacks, 32
Stings, 118, 121, 146, 152, 162
Stomates, 44, 102, 226
Streams, 11, 15, 29, 32, 42, 83, 146
Submergents, 46, **49**–50, 52–53, 89, 102,
  105, **150**, 226 (see also specific
  submergents)

Succession, 28–55, 226
  early stages of, 28–29, **32–33**
  late stages of, 28–29, **51–53**, 55
  mature stages of, 28, **34–35**, 41, 44, 46,
    50–51
Suffocation, 33, 95, 178
Summer, 11, 57, 66–72, 79–83, 89, 95,
  97, 107
  aging ponds in, 52–53
  daily cycle during, 76–77, 79
  food lost and restored in, 82–83, 88
Sunfish, 32–33, 46, 62, 66–67, 72, 81,
  130
Sunlight, 48, 57, 80, 82, 111, 185, 197
  autumn, 89
  as heat source, 79, 97, 110
  and photosynthesis, 23, 153, 177–178
  pollution and, 106
  spring, 58, 62, 65
Surface film, 20, 41, 123–125, 170–176
  inhabitants of, 121, 149, **171–175**
  as trap, 83, 175
Surface tension, 119, 170, 173, 226
Swamps, 10–11, 35
Swimmerets, 129
Swimming, 108–110, 114–123, 156, 159
  of insects, 108, 116, 118–119
  of plants, 185, 191
Symbiosis, 152–153, 159, 182, 226

Tadpoles, **24**, 67–71, 82–83, 117, 134,
  146, 149
Teals, **134**
Temperature (see Atmosphere, temperature
  of; Body temperature; Water
  temperature)
Tentacles, **120–121**, 152, 162–**163**
Terrapins (see Turtles)
Theca, **88**
Third-order consumers, 24, 26–27
Thrips, 168
Tiger beetles, **196–197**
Toads, 67–69, 140
Top feeders, 170–171
Tracheal system, 124, 127–128, 226
Trachelomonas (flagellates), 185
Transparent water, 46, **48–49**, 66,
  106–107
Traps, 83, 164, 175, 181
Trees, 34, 52, 55, 144
Trichodina (peritrichs), **152**
Tubercles, 197
Tubifex worms, 79, 179
Tundra ponds, **30**
Turtles, 41, 46, 89, 109–113, **166**, 213
  breathing of, 123–124
  eyes of, 79, **113**, 156
  in food chains, 25–26, 28, 134, 138,
    140–141, 145, 150
  as hosts, 115, 152, 156
  limnetic, 20
  migration of, 52, 80
  as scavengers, 159, 179

Unionicola (water mites), 128

Vision (see Eyes)
Volvox (flagellates), **190–191**
Vorticella (peritrichs), 162–**163**, **169**

Waste products, 82–83, 88, 128, 146,
  179–180
Water beetles, 41, 59, 65–66, 77, 79,
  115–118
  breathing of, 124, 127
  breeding of, 168
  on surface film, **170–171**, 174–175
  winter activity of, 95
Water boatmen, 51, 65–66, 79, **118**–119,
  125, 127, 153

Water bugs, 41, **146**
Water density, 95, 97, 155
Water fleas, 25, 59, 77, 121, 153, 162,
  175, **193–194**
  estivation of, 80
  as filter feeders, 182
  in food chains, 149, 151
  habitat of, 171
  winter population of, 95, 97
Water lilies, **22**, **45**, 76, **80**, 134–**135**
  in autumn, 88–89
  flowers of, **104**–105
  as microhabitats, 166–169
  mucus on, 103
  yellow (see Spatterdock)
  zone of, 44
Water milfoil, 46–**47**, **103**, 105
Water mites, 70, 100, 128, 152–156,
  168–**169**, 194
Water net, **184–185**
Water pollution, 29, 106, 181, 184
Water pressure, 127
Water scorpions, 123, 127, 153–**154**, 168,
  170
Water shield, 44, 134, 166
Water smartweed, 88
Water snakes (see Snakes)
Water sources, 29
Water striders, 41, 59, **150**, **170**, **172–173**,
  175
Water temperature, 14–15, 25, 42, 66–67,
  89, 212
  layers of, 11, 66
  oxygen supply and, 79, 131
  springtime, 58, 62
  wintertime, 95, 97
Water velvet, 106
Water willow (see Loosestrife)
Waterfowl, 16, 34, 72, 76, 105, 140
  diving, 50, **132–133**
  in otters' diets, 108–109
  as removers of pond food, 82–83
  zones frequented by, 50
  (See also specific waterfowl)
Watermeal, **44**, 59, 105–106, 162, 171
Water-regulating devices, 100–101
Water-repellent hairs, 125, 127, 149,
  171–172
Waterweed, **46**, 89, 103–104, 134
Whirligig beetles, 41, **116**, 124, **169**–170,
  174–175
Wind, 44, 46, 66, 88–89, 105, 171, 177
Winter, 19, 57–59, 86, 88–89, 95, 97,
  103, 105, 131, 146, 191
Worms, 9, 20, 32, 52, 108, 146, **150**
  burrowing, 79, 95, 178–179, 196–197
  in food chains, 147, 149–150
  phoresis of, 152
  stem habitat of, 168–**169**
  (See also Flatworms)

Yellow-brown algae, 106
Yellowstone National Park (Wyo., Mont.,
  Idaho), 205
Yosemite National Park (Calif.), 205

Zonation (see Animals, zonation of;
  Bottoms; Ecotones; Limnetic zone;
  Littoral zone; Plants, zonation of)
Zooplankton, 25–26, 53, 59, 88–89, 121,
  183, 192–195, 226
  filter feeding and, 181–182
  insects in, 119, 149
  as light shield, 177
  nighttime activity of, 77, 149
  as primary consumers, 133
  summer growth of, 80
  (See also Copepods; Flagellates;
    Larvae; Rotifers; Water fleas)

232